Josiah Foster Flagg

Plastics and Plastic Filling

As Pertaining to the Filling of All Cavities of Decay in Teeth below Medium in Structure and to Difficult and Inaccessible Cavities in Teeth of All Grades of Structure. Fourth Edition

Josiah Foster Flagg

Plastics and Plastic Filling
As Pertaining to the Filling of All Cavities of Decay in Teeth below Medium in Structure and to Difficult and Inaccessible Cavities in Teeth of All Grades of Structure. Fourth Edition

ISBN/EAN: 9783337811600

Printed in Europe, USA, Canada, Australia, Japan

Cover: Foto ©berggeist007 / pixelio.de

More available books at **www.hansebooks.com**

"This is a comprehensive and well-written book on the subject of which it treats. The author's professional standing and large experience justify the publication of his views with regard to the employment of plastic materials in the treatment of decay occurring in teeth. His remarks concerning the employment of gold as a stopping in decay of teeth which are below the average development of structure, or in which, by the inaccessible position of the cavity, a perfect gold stopping cannot be insured, are well worthy of careful perusal. The author states emphatically that many teeth are lost owing to the injudicious use of gold. Such a statement, coming from one who has been nurtured in the land of 'gold stopping,' would scarcely have been made without mature observation and calm reflection. We are certainly much inclined to his views, and are quite in accord with him in maintaining that the chief duty of the dental surgeon should be to preserve the dental tissues, rather than to inquire at how early a date his patient may need those artificial substitutes which at the best are but imitations of their natural predecessors. The work is well arranged and admirably illustrated."—*London Lancet.*

"This volume is an octavo, printed on excellent paper, with wide margins, in a good, clear-faced type, and is well and tastefully bound, making altogether a very fine appearance and an exceptionally agreeable book to read. . . . Whatever may be thought of Dr. Flagg's theories or practice, he has evidently made an honest and earnest effort to give the benefit of his experience to the profession. The teaching is so explicit that none need complain that there is any want of definiteness in it."—*Dental Cosmos*, April, 1881.

"A less courageous man than Dr. Flagg would have hesitated before presenting to the profession a book of this character—not, however, that the character of the book is bad. Dr. Flagg has taken the plastic bull by the horns, and forced said bovine to the front, and sustained him in that position by a remarkably clear and precisely written volume. . . . If plastic materials are used at all, they should be prepared and inserted in the most precise and careful manner, and Dr. Flagg characteristically shows how this can be done."—*Dental Advertiser.*

"Dr. Flagg's treatise has great literary merit. It is logical, forcible, and so clear in its statements that scientific facts, theories, and methods are made plain to unscientific readers. But its greatest merit is that it illustrates the second great advance in the progress of dentistry, and that it gives expression to principles and methods of investigation, which, if followed in the future with the same intelligent enthusiasm, must produce very important results."—*Philadelphia Inquirer.*

PLASTICS

AND

PLASTIC FILLING;

AS PERTAINING TO THE

FILLING OF ALL CAVITIES OF DECAY IN TEETH
BELOW MEDIUM IN STRUCTURE,

AND TO

DIFFICULT AND INACCESSIBLE CAVITIES
IN TEETH OF ALL GRADES OF STRUCTURE.

BY

J. FOSTER FLAGG, D.D.S.,

PROFESSOR OF DENTAL PATHOLOGY AND THERAPEUTICS IN
PHILADELPHIA DENTAL COLLEGE.

WITH ILLUSTRATIONS.

FOURTH EDITION, REVISED.

PHILADELPHIA:
1891.

COPYRIGHT.
J. FOSTER FLAGG.
1890.

ALL RIGHTS RESERVED.

STEREOTYPED BY J. FAGAN & SON, PHILADELPHIA.
PRINTED BY SHERMAN & CO.

TO

THE SCIENTIFIC, METALLURGIC, AND DENTAL SECTIONS

OF THE

"NEW DEPARTURE CORPS,"

THIS MONOGRAPH

IS INSCRIBED, AS A TESTIMONIAL

OF INDIVIDUAL INDEBTEDNESS,

OF SINCERE RESPECT,

AND OF KIND REGARD.

PREFATORY.

IN explanation of the occasionally desultory arrangement of the matter in these pages, and of some of the peculiarities of composition, I would say that the "Articles" were originally intended for magazine contributions, but, in evolution, the statements which I decided to make, the information which I desired to give, and the deductions which would necessarily have to be drawn, were such as I could not ask any magazine of extensive circulation, with extensive business interests, to publish. I have therefore concluded to offer them in this form that I might, for the good of my profession — as I hope and as I trust — disseminate, in all its integrity, the results of that work which has been done by the "New Departure Corps" under its motto, " Truth, without fear and without favor."

J. FOSTER FLAGG.

PREFACE TO THIRD EDITION.

RADICAL changes in amalgam alloys and their utilization; decided advance in work pertaining to both manufacture and manipulation of G. P. Stopping; modifications in connection with both zinc-phosphate and zinc-chloride; need for more specific directions regarding manipulation of zinc-chloride, and for many stereotypal corrections—which have had to be made in ink in previous editions—have compelled such extended revision of matter for this Third Edition of "Plastics and Plastic Filling," as to have made of it a labor which has been exceeded only by the long and diversified experimentation that has eventuated in the results now presented.

It is my earnest desire that this "Third" effort may teach means and methods for tooth-salvation which shall notably exceed even those of former editions, and with this hope it is offered to the practitioners of the Profession of Dentistry.

CONTENTS.

INTRODUCTORY.

Difference in basal principles which exists between the Schools of Gold and Plastic Dentistry 11–18

ARTICLE I.
PLASTIC FILLING.

History of the work upon the "Systematic Development of Plastic Filling" 19–23

ARTICLE II.
AMALGAM.

Silver Paste; Royal *Mineral* Succedaneum; Action of the American Society of Dental Surgeons; Dissolution of that Association . . . 24–29

ARTICLE III.
AMALGAM.—Continued.

Facts worthy of consideration; Prof. Elisha Townsend's sanction of amalgam; His acceptance and advocacy of proposed improvements in alloy and making of amalgam; Action of the representative men of the profession; Professor Townsend's "recantation;" The three basal positions in connection with amalgam which decided me in continuing work, notwithstanding my leader's "recantation;" Conclusions reached by five years of experiment and observation 29–36

ARTICLE IV.
AMALGAM.—Continued.

From 1861 to 1875; Change of formula; Adoption of "lining;" December, 1874, Meeting of New York Odontological Society; Metallurgic association with Messrs. Du Bois and Eckfeldt; Comments upon paper of Professor Hitchcock; Comparison of Analyses of Amalgam Alloys; Remarks on "Gold and Platina" alloys; Remarks on "advanced collegiate instruction" 37–42

ARTICLE V.
ATTRIBUTES OF METALS USED FOR AMALGAM ALLOYS.

Definitions of "alloy" and "amalgam;" Theory of the formation of amalgams; Metals used in amalgam alloys; Silver; Tin; Copper; Gold; Antimony; Zinc; Cadmium; Platinum 43–60

ARTICLE VI.
THE MAKING OF AMALGAM ALLOYS.

"Accepted" methods given by different instructors; The completely different method of the "New Departure" metal workers; Cutting of alloy into "grains," "filings," or "a kind of powder;" Inferiority of "fresh" cut alloy; Superiority of "aged" alloy filings; *Qualitative Testing of Amalgam Alloys* 61–66

ARTICLE VII.
TESTS FOR AMALGAM.

Quality test; Shrinkage test; Color test; Edge-strength test; Tooth-conserving test 67–80

ARTICLE VIII.
PREPARATION OF CAVITIES.

Comparison of preparations for gold, with preparations for amalgam; Effect of amalgam upon contiguous dentine 80–84

ARTICLE IX.
THE MAKING OF AMALGAM.

Former methods; Remarks on alcoholic "washing;" Mortar-make; Weighing for proportions; Ground-glass mortar and pestle; The method of rubbing; Method of kneading; The forming of a "button;" 84–92

ARTICLE X.
INSTRUMENTS FOR THE INSERTION OF AMALGAM FILLINGS.

Description of instruments; Round ends; Flat ends; Trimmers or Separators; Spatula; Elliott's "Loadstone" amalgam-carrier and plugger; Elliott's amalgam carrier; Chase's amalgam carrier; Fry's amalgam carrier; Mullett's "Amalgam Director;" Curious and useless appliances 92–99

ARTICLE XI.
THE INSERTION OF AMALGAM FILLINGS.

Necessity for *acquirement of manipulation* by nearly every dentist, from the marked difference in physical characteristics between "accepted" amalgams, and those offered by the "New Departure;" Present division of amalgams; Submarine; Contour; Front Tooth; Facing; Copper; Coin Amalgam, Insertion of amalgam mass; Wafering; Shaping; Smoothing; Finishing; Whitening; Subsequent burnishing, if indicated; Remarks upon burnishing the edges of partially-set amalgam fillings 99–107

ARTICLE XII.

GENERAL CONSIDERATIONS PERTAINING TO AMALGAM.

1. Local effects; 2. Systemic effects; 3. Possibilities; FIRST. *Local Effects.*—Discoloration of filling and tooth; Induction of galvanic electricity by contact with other metal; Shock; Metallic salivation; Bad taste; Irritation of fauces, throat, and larynx; Pulp-devitalization; Periodontitis; Alveolar abscess; Exostosis; Necrosis.

SECOND. *Systemic Effects.*—Proceedings of Pennsylvania Association of Dental Surgeons, April, 1861; Testimony of members adverse to mercurial ptyalism, mercurial necrosis, etc.; My own opinion regarding mercurialization from amalgam fillings.

THIRD. *Possibilities.*—Possibilities of amalgam are greater than those of *all other filling materials* combined; Comparison of the working of gold with that of amalgam; Ordinary cavities; Large cavities; Enormous cavities; Entire crowns; Attaching gold crowns by amalgam; Replacing natural crowns; Attaching natural crowns; Pivoting teeth; Bonwill pivot; Gates Pivot; Flagg pivot; "Guarding" or repairing gold fillings, and refilling cavities from which gold fillings have been lost, in teeth which still contain gold fillings; Bi-metallic fillings; Therapeutic value of these; Amalgam teeth on gold plates; Fractures of teeth from fillings either of gold or amalgam; Split teeth; Perforated teeth; Loose and divided roots; The dividing line between the impossibilities of gold work and the continued easy possibilities of amalgam work; The *comfort from plastics* 107–141

ARTICLE XIII.

GUTTA-PERCHA.

Hill's stopping; Gutta-percha and Oxide of Zinc Stoppings; Leakage of gutta-percha fillings; "Low-heat" gutta-percha; "Medium heat;" "High heat;" Red base-plate; Oil-pad; Modes of heating gutta-percha filling materials; Methods of heating instruments; Considerations pertaining to preparation of cavities for gutta-percha fillings; The introduction of fillings; Finishing fillings; Tests for gutta-percha stopping; Heat-test; Fire-test; The making of gutta-percha stopping; Durability of gutta-percha fillings; Cavities in which the use of gutta-percha alone is indicated; The varied uses of gutta percha stoppings 141–150

ARTICLE XIV.

OXY-CHLORIDE OF ZINC.

Sorel cement; Uses of oxy-chloride of zinc; "Talcing" fillings; Oxy-chloride not a filling material; Oxy-chloride a complete "liner;" Failure of oxy-chloride from attrition; Failure of oxy-chloride from cervical disintegration; The making of oxy-chloride powder; The making of the fluid; The mixing of the materials for filling; Method of lining cavities; Results attained by the use of oxy-chloride; Remarks in relation to "combination" fillings; "Whitening" teeth by the use of oxy-chloride 151–160

ARTICLE XV.
OXY-SULPHATE OF ZINC.

Not a filling material, but eminently useful as an adjunct to plastic fillings; The making of oxy-sulphate powder; The making of oxy-sulphate fluid; Oxy-sulphate as a pulp-capper; Its mixing; Capping by spatula placing; Capping by pellet placing 160–162

ARTICLE XVI.
ZINC-PHOSPHATE.

Ostermann's formula; The *zinc-phosphates* of the present; The *oxy-phosphates*; The making of zinc-phosphate powder; Tests for distinguishing it from oxy-phosphate powder (calcined oxide of zinc); Remarks on oxy-phosphate and zinc-phosphate fluids, syrups, and crystals; Mixing zinc-phosphate; Working tests for grading a "good" zinc-phosphate material; Working tests for grading a "poor" zinc-phosphate; Directions for using zinc-phosphate; Varnish recipes; Final manipulative suggestions; Conclusions as to value of zinc-phosphate cement 162–174

ARTICLE XVII.
TEMPORARY STOPPING.

Directions for making 174

ARTICLE XVIII.
TECHNICALITIES.

Ageing 175	Mixing 191
Bulging 176	Pelleting; Rubbing . . . 193
Buffering; Capping; Cold-Soldering 177	Setting 195
	Softening 197
Crevicing 179	Shrinkage 198
Domeing; Facing 180	Tapping; Testing . . . 199
Frotting 182	Trimming 202
Guarding 183	Trunnioning; Wafering . 203
Heating 189	Washing; Weighing . . 204
Lining 190	Whitening 206

CONCLUSION 207
APPENDIX 208
CONCLUSION TO THIRD EDITION 211

Illustrations.

	PAGE
Monogram, N. D. C.	iii
Mixer	opposite 65
Index Micrometer	" 74
"Shrinkage" Microscope	" 75
Edge-Strength Tester and Ingot Matrix	" 77
Amalgam Balance	" 89
Mortar, Pestle, and Mercury Holder	" 90
Amalgam Instruments	" 93
Amalgam Carriers, etc.	" 98
Flagg's Ring-Pivot	" 128
Gutta-Percha Warmer and Tool-Heater	" 144
Gutta-Percha Instruments	" 146
Gutta-Percha Heat-Tester	" 147
Frotting Tester	" 183
Wafering Pliers	" 204

PLASTICS

AND

PLASTIC FILLING.

INTRODUCTORY.

BEFORE entering upon considerations pertaining to the historic, analytic, utilizing, and manipulative detail of our subject, I feel desirous that the thoughts of my readers should be so directed as that there shall be, at once, a recognition of the great fundamental differences in basal principles and practice which exist between the Schools of Gold and Plastic Dentistry.

As facts in relation to present methods and results are accumulating, I am the more impressed with the growing need for the uprooting of that tendency in dental teachings and practice which has, thus far, resulted in such waste of time, energy, strength, money, *and teeth*, and which still, in some sort, holds sway at the gatherings of the "wise men."

The idea of pounding piece after piece of gold into some inaccessible pocket, far below the gum, in an "annex" to a cavity almost into the pulp, on the distal face of a lower molar of poor structure, and calling it "the highest attainment of first-class dentistry," is, to me, something incomprehensible! It is but trifling with the dignity, the broad ability, the glorious possibility, of our profession. Nor is this an overdrawn picture; it is precisely what is yet done — done "by and with the advice and consent of the Senate," and it is truly an imposition upon patients and upon the younger dentists, inflicted in the name of science and of art.

The depth and the breadth of the work to which I have been called impresses me most earnestly as I view more and more clearly the vast difference between "Gold-work," regarding that as one school of dentistry, and "Plastics," regarding that as another school of dentistry.

Outwardly, the two schools are very different even to the most casual observer, but in proportion as one is educated in both, this external difference becomes even more tangibly apparent; the materials are different; the appliances are different; the methods are different; the instruments are different; the attempts are different; the possibilities are different, and *the results are different.*

The time has gone by when to this last assertion the "gold-workers" can say, with a palpably implied tone of superiority, "Yes, the results are, indeed, widely different," for *the people* are trumpeting loudly their stories of heart-sickening failures and wholesale loss of teeth. The mouths of a fearful proportion of this generation reveal only too glaringly the peculiar curve of "porcelain teeth mounted on vulcanite," and *the demand* is for something *far different* from that which has heretofore been given them as "dentistry."

But great as are the *external* differences between the "gold-workers'" practice of dentistry and its practice from the "plastic" standpoint, the *internal* differences are vastly greater. The *ideas* of the two kinds of practice flow in two channels, than which no two could be more distinct. The *thoughts* which govern work in the two kinds of practice are constantly almost diametrically antagonistic, and are *never*, in any degree, sympathetic.

The minutiæ of "plastics" is an unknown thing to the gold-worker. It consists of special knowledge in regard to much which is only known to him as a mass of vague generalities; while the "ways and means" of the worker in plastics are viewed askance by the gold-worker as a series of shiftless devices promotive of slovenly results.

The minutiæ of "gold-work" is utterly ignored by the worker of "plastics," and for such "ways and means" as he sees the gold-worker resort to in "difficult cases," the worker of plastics has a shuddering horror!

The first view of *any case* in practice, as seen from the two standpoints, is provocative of such different impressions upon the two practitioners as renders it impossible that the one should have any conception of the effect made upon the other; and only he who is conversant with both schools can, in full degree, appreciate this.

Take, for instance, in a perfect arch, a lower second molar; soft structure; largely decayed distally and buccally; the decay extending deep below the gum in both directions and running out to feather-edges; with frail, overhanging cavity walls mesially and lingually; the pulp almost exposed; the patient, a lady of nervo-bilious temperament, middle-aged, a sufferer from congestion of the liver, and overtaxed nervously.

What is the first thought of the "gold-worker"? It is, How shall I get a *good, solid, gold filling* into this cavity?

What is the first thought of the "worker in plastics"? It is, What is it *best* that I should fill this cavity with?

Next comes the cavity preparation. The thoughts of the gold-worker are concentrated upon the making of free access for the introduction and packing of his gold; upon the securing of strong, smooth edges; upon the making of retaining points; upon the removal of all decay, except such as would endanger the pulp, that his gold may have a solid foundation to rest upon, and that he may be able to introduce a filling which will not leak; upon the possibility of "capping,"—thinking of "conduction" of filling material,—and upon the squaring and the grooving of the walls at feather-edges.

The thoughts of the worker in "plastics" are upon the conservation of enamel; the spheroiding of cavity contour; the conservation of decay,—within the bounds of filling integrity,—hardly bestowing a thought upon the pulp, and certainly no thought upon pulp irritation from "conduction;" choosing, mentally, a filling possessed of "edge-strength" in view of feather-edge to cavity; selecting his most trustworthy "submarine" in view of depth of cavity below the gum; projecting a "combination filling" which shall be composed of four metal amalgam — mercury, silver, tin, and copper; five metal amalgam —

mercury, silver, tin, copper, and gold. Good gutta-perchas,—red or white "low heat" and white "high heat,"—each filling material used in its proper place for the attaining of some specific result; and thus, according to *his* ideas, best securing against recurrence of decay; best affording good maintenance of color and contour for resistance in mastication; and best securing that uninterrupted pulp action which shall not only preserve the integrity of the organ, but shall eventuate in re-calcification of the remaining substratum of decalcified dentine.

Next follows the introduction of the filling. Here the serious labor of the gold-worker begins. He is an earnest, conscientious man; he is an eminently skilful manipulator; he glories in his work; his whole thought *must* now be concentrated upon the introduction of his filling; any interference with this precludes the possibility of success.

As a necessary prelude, the decayed tooth and the adjoining teeth must be placed "under rubber." The work of adjusting the rubber-dam, in such a case as this, is one requiring much knowledge, much patience, and much dexterity; knowledge of fitness and quality of rubber-dam, knowledge of punches and their using, knowledge of clamps and clamp-forceps; patience under difficulty, patience oftentimes under repeated tearing and slipping of the dam, patience under remonstrance at infliction, patience under failure until success is attained; dexterity in forcing clamps below the gum, dexterity in passing rubber over clamps and over and between the teeth, dexterity in placing and securing retaining ligatures.

All this work is very exhausting both to patient and operator; and it is, therefore, under conditions trying alike to mind and body that the difficult task of the introduction of the filling is, at last, *commenced*.

Then comes a work of *hours*—two, three, it may be *four*—in such a cavity as this. Whether it be of pellets, rope, or cylinders; whether it be of soft gold or cohesive; whether it be by hand-pressure or with mallet,—automatic, hand, engine, or electric,—it is, by any means, in any way, *a famous piece of work*. It is a work which can be *best* done by not more than one worker in a thousand; a work which can be *well* done by not more than one worker in a hundred; a work which is not *reasonably well done* by more than one worker in ten; and yet a work which is *attempted in gold-working dentistry by nine workmen out of ten.*

And during all the progress of this work, what are the *thoughts* of the gold-worker? First, his thoughts are on the choosing of his gold. Shall it be "Abbey's," "Globe," or "Ashmead's"? Shall it be "Morgan's" or "Watt's Crystal"? Next his thoughts are on its preparation. He revels in the beautiful. All his ideas are æsthetic. This is the "power" of *such* gold work; its strength is in its *beauty*, and in the beauty of its surroundings, and it must be recognized as possessing it; so he deftly handles his "crystal" as he picks off little morsels, or folds his golden ribbons and cuts them into pieces, or inrolls pellets, or twists sheets into ropes, all laid on skins of kid or dropped on velvet cushions.

This being done, he thinks of final "warm-air" drying; and next on the successful filling of the first retaining point; and then on the successful filling of the second retaining point. Then his thoughts are upon the connecting of these two fillings by a "first layer" of gold; and all this is at the "vulnerable spot," at the disto-cervical edge. Not a piece must move; not a piece must miss its "weld;" not a piece but must be placed with that *tension of accuracy* which is so well known to those of us who have been repeatedly prostrated under its depressing influence, or *the work will "fail from defective manipulation."*

It is just here that relief is sought by the gold-worker in the possibilities of soft foil pellets and hand pressure; and it is upon the respective merits of soft gold and cohesive, in this connection, that discussion has been going on for thirty years; the experiences of to-day are those of a quarter of a century since; the arguments and assertions of *then* are those of *now*. The subject has been hammered at until it is utterly attenuated; and in its gossamer-like thinness it is hammered at with undiminished vehemence; so *it is evident* that the gold-worker regards it as worthy of *much thought.*

The records of all the meetings of the American Dental Association prove that no conclusion upon this point has been reached, so that *nothing definite* can here be thought of by the gold-worker; but by either mode the filling gradually grows; pellet is added to pellet, or fold after fold of rope is inserted, or piece to piece is united, until finally the thought of the gold-worker is, *the gold is all in.*

Next come the thoughts of finishing; and this is also a prolonged task. Thoughts of files and burrs and stones; thoughts of corundum tape, pumice and tutty powder; thoughts of burnishers, flat, round, and ball; thoughts of the beautiful, lustrous soap polish.

And in this work, as "hand, responsive to the thought," develops, step by step, the "elegant conception," at last 'tis finished; and some of the final thoughts are, truthfully, that "it is perfectly magnificent;" that "it is evidence of exceeding skill;" that "it is artistic."

To all this the worker in plastics says, Amen! But others of the final thoughts are, that "it is the *best* that could be done;" that "it is a proof of *superiority* as a *dentist*;" that "it is the *most durable* as a preserver of the tooth."

To these thoughts the worker in plastics does *not* say, Amen!

He thinks, "I would not like to have that kind of thing done to *my* tooth." "It certainly does prove that the gentleman is a superior tooth-jeweller; but I do not yet exactly recognize the *dental superiority.*" And *he* next thinks of the *teeth of that kind* which *he has seen* in which fillings of amalgam and gutta-percha, so COMFORTABLY INTRODUCED, have done *ten years* of service, and bid fair to do years of service yet, in which just such beautifully polished jewels had failed in from *three* to *eight* years; and if the gutta-percha is worn and cupped, and if the amalgam is discolored, *he* feels toward them as he would toward a stalwart old negro who had carried him safely over a roaring river,—he views leniently the scars and furrows wrought by time and exposure, and even forgives the fellow for having *a black skin!*

As with the gold-worker, so with the worker in plastics. After the preparation of the cavity, next comes the introduction of the filling; but with what different feelings and thoughts does he come to the work. Selecting one of his thinnest separating-slides he bends it to the desired shape, adapting it as a broad clasp to the wisdom-tooth, with the outspoken thought, "it will be less inflictive to withdraw a separating-slide than to cut a separation between the filling and the adjoining tooth with the cavity so deep under the gum;" and then, drying the cavity with bibulous paper, prepares for the introducing of his material by simply pressing into the cavity pellets of cotton moistened with oil of cloves, with just sufficient force to push aside the gum; he next selects his "submarine" alloy, with thoughts of the requirements and knowledge of its composition and proportions; he recognizes its liability to discolor, but thinks of the tooth-saving effect of discoloration; he recognizes the difficulties with which he has to contend, and thinks cheerfully, both for himself and the patient, of the facility with which, by his mode of practice, he will overcome them. And with these thoughts he has weighed and made his material; and now, removing the cotton pellets and again drying the cavity, proceeds dexterously with the insertion of the lower stratum of his filling. With the "rough trimming" of his work, his thought is "with what ease and celerity has this so-called difficult operation been transformed into one of perfect simplicity;" and with the remark that "it will now have to wait for fifteen or twenty minutes," he proceeds with another preparation or, for the few minutes, dismisses the patient. In due time he removes his "slide," does what little finishing trimming may be required, and then proceeds to place the tooth "under rubber." He has no thoughts of clamps or clamp-forceps; he has but little, if any, thought of ligatures; *he has removed everything* that makes the placing of the rubber a trying, difficult, and painful thing to the *gold-worker* and *his* patient; and thus, having punched three or four holes, his beginning, instead of being the forcing of a clamp far below the gum,—an operation which must be *felt* ONCE to be appreciated,—is the easy slipping of the dam over the bicuspids, and the almost equally easy successive damming of the molars; he thinks, with much inward satisfaction, that the rubber does not have to go below the gum, and that, *if* a ligature is needed, it can be placed with comfort to the patient.

Then his work may be considered practically done; for, with the cavity perfectly dry, the manipulation of "plastics" is sure and easy; the low-heat gutta-percha adheres to the softened dentine, covering the pulp with a degree of heat which is unnoticed even by that sensitive organism; the quick-setting alloy, of good color test, is rapidly and effectively placed in position without a disturbing thought of "flaking," "missing weld," or "loosening of the filling at the vulnerable spot," and the final finishing on the buccal face with the "high" gutta-percha completes a filling which to the worker in plastics suggests a long train of satisfied thoughts as to its acceptable introduction; its happy meeting of varied requirements; its presentable, tooth-like appearance, and its probable value as a durable, pulp-saving, tooth-saving, comfort-giving operation.

ARTICLE I.

PLASTIC FILLING.

PREMISING that the decided advantages which would accrue from a gradually increasing employment of plastic filling materials in connection with efforts for the saving of teeth, *below medium in quality*, are fast becoming recognized by a very large portion of the practitioners of dentistry, and believing that their successful utilization depends as much upon a *scientific adaptation of means to ends*, as does the successful utilization of gold depend upon *manipulative ability*, I purpose offering some suggestions which, while they may not fully educate to the use of "plastics" from the high analytical standpoint that must entirely govern this when based upon future collegiate instruction, will, nevertheless, give such aid to those who desire to experiment in this direction as will enable them to produce results which will be eminently satisfactory alike to patient and operator, and this, too, in the very cases where even the most skilful manipulators in gold fail ignobly.

To this end I shall first discuss the indications which point to the employment of any *single* plastic material for the making of a filling, and then pass to the consideration of those *combinations* which are the *essentials* to the perfect development of that system of tooth-salvation which is based upon the electro-chemical theory of Dr. S. B. Palmer, and which attributes the failure in operations *mainly* to "incompatibility of filling material with tooth-bone."

It is now just thirty-five years since I commenced the work of "Systematic Development of Plastic Filling." It has been, with me, a "labor of love," one in which I engaged under the favorable auspices of hearty encouragement from professional elders for whom I had the highest respect; whose attainments as practitioners were unquestioned, and whose allegiance to dentistry was undoubted.

It has been a long, and sometimes tedious, path which I have trodden, but the interest has been ever increasing, and the gleanings from the way-side have proven to me, and to thousands of my fellow-creatures, productive of an ample harvest of mutually satisfactory results.

During the first six years, progress was naturally slow, but at the expiration of that period I had obtained such data, as that the prosecution of experiments seemed not only warrantable, but to be *demanded*.

As I was early educated to the value of "a record" I desired, as a preliminary to long continued, further work, that the conclusions which had been obtained should be *recorded*, that they could be referred to, be the outcome what it might.

To this end, I proposed, as the subject for discussion at the meeting of the Pennsylvania Association of Dental Surgeons, held March 12, 1861, "the consideration of the so-called *osteo* plastic materials for filling teeth." Although the subject was not so stated, the discussion was, under direction, confined to the then new "oxy-chloride of zinc." As *I* had "proposed" for that meeting, I asked my friend Dr. C. Newlin Pierce to propose "amalgam" as the subject for the next meeting, April, telling him that I desired particularly to obtain a general expression of opinion in regard to that material, and also wished to place myself upon record for an especial purpose.

Reference to the discussion of that subject as found in the *Cosmos* for May, 1861, page 548, will show that I had not only the results of my own "*six years*" of observation, but the markedly favorable views of the members toward amalgam, to sustain me in a continued series of investigations in connection even with that hitherto questionable compound.

This was, to me, a matter for congratulation, as I was then about to commence a gradual increase in relative proportion of plastic fillings to those of gold, and it became important that this increase should be definite in materials as well as in degree. I therefore decided upon amalgam, gutta-percha stopping, and oxy-chloride of zinc for materials, and upon six *per centum* of fillings as an amount of annual substitution of plastics for gold which would permit of careful observation and of healthful progress.

This, of course, could not be done *positively*, but it was done *so approximately*, as that at the close of the ninth year I found myself filling all cavities in teeth *below medium in structure* with plastic materials.

By this time the demands for services had reached a point which permitted of *selection of cases* upon my part, and I resolved to confine my work thenceforth, as much as possible, to the *saving of soft teeth*, a labor which I *then recognized* would eventuate in the *exclusion of gold as a filling material*.

It can readily be imagined that I viewed this result as a matter of serious import, for I felt that its outgrowth must naturally be an uncompromising attack upon the time-honored and fondly cherished articles of belief that "gold is the *best* material for filling teeth," and that "eminent skill in working gold is the basal requirement for superiority as a dentist;" but even for this I had been collaterally prepared, largely through the instrumentality of my cherished friend and able instructor, Prof. Robert Arthur.

NOTE.—It is a source of the greatest comfort to me to have the written proofs of the warm friendship which existed between Prof. Arthur and myself even to the period of his last illness, and to be able to say that he smiled approvingly upon the picture of himself, which he had presented to me, when he saw it in my office as the *first* in the "Gallery of Dental Heretics."

I had already enjoyed the opportunity for observing the excellent results of many years of practice based upon his teaching in regard to the leaving of decay in cavities, for the protection of the pulp. I had heard the violent denunciation with which this teaching had been assailed, and had seen the coldness with which the "eminent" men of 1860 had turned their faces from this "disreputable" practice; and I had lived to hear this innovation taught, as universally accepted, and to see that the vast majority of the virulent opponents had been forced to practise upon this teaching.

This naturally gave me faith to believe that other tenets, *if built upon the sand*, would fall, and that other teachings, *if founded on a rock*, would stand.

Time and experiment would establish truth, and in no other way could it be established; therefore, from this period, I declined all new patients with teeth above medium in structure, and more closely concentrated my experiments upon such teeth as are, even now, usually condemned to extraction.

In seven years more I found myself with a practice which was and is, to say the least, *peculiar*.

It is peculiar, in that I seldom see a denture which is above average in quality.

It is peculiar, in that I have many hundreds of patients for whom I have saved, for many years, the remains of dentures which had been almost lost under excellent operators with gold.

It has been freely suggested to me that it is easy to save teeth when all those predisposed to decay have been extracted! and that it is easier to save teeth when arches are so broken as to permit a natural separation of the teeth! and that persons' *constitutions are apt to change*, and thus the teeth are less liable to decay! but it has seemed to me strange that so many patients have come *just when every tooth which was predisposed to decay had been extracted*, and had thus placed under my charge *no teeth except such as were not liable to decay!* And again, I have wondered why, when arches were quite sufficiently broken to permit the natural *saving separation* of the teeth, the loss should be continued until there were not teeth enough left to permit of natural mastication! and yet again, to me it seemed the most remarkable of all, that every patient's constitution should change immediately upon placing the teeth under the influence of plastic filling materials!

It is peculiar, from the fact that I am able, with such dentures, selected for their frailty, to meet all indications without resorting to "artificial work." By this term I mean to refer to the introduction of sets of teeth upon plates, either partial or complete, and do not include as "artificial work" the grafting of porcelain crowns, the insertion of plate teeth as faces to "pivotings" or the introduction of porcelain as edges, corners, or as "mosaic" fillings; all this work, I think, is properly taught as "operative," and I would be understood as so regarding it.

During the past eight years, the requirements of all my patients have not reached an average of one piece of artificial work *per annum*.

It is peculiar, in that I have no demand for gold as a filling material. When I had reached this point, it was with little or no surprise that I beheld this peculiar result; the demands for gold had been steadily diminishing for years; as successive exigencies had arisen, each difficulty had been satisfactorily surmounted, and each surmounting had opened the way for easier conquering of greater difficulties; in this work the claims of gold became less obtrusive, until they were finally lost sight of completely.

With teeth which had been filled and refilled with gold until the expenditure had been enormous; until the repetition of infliction had been such that the mere thoughts of dental visits had become intolerable; until the gradual increase in frequency of extraction had become an accepted institution; until all hopes of comfort or tooth-salvation had been lost—what was there to offer in continuance of such practice? .

With those who had honestly and earnestly made diligent inquiry for "the best,"—as *best* after *best* had failed to meet their requirements,—and who had availed themselves of services, the remaining results of which were evidences of marvellous skill, patience, and endurance, and with whom all had proven but a mockery, what was there to offer in continuance of such efforts?

With those who were thoroughly dissatisfied with what had been done for them, and who had ample and unanswerable reasons for their views of "first-class" dentistry, in what manner would any proposition be received other than for something entirely different in dental work from anything which they had previously experienced?

Under such circumstances, what more positively unavoidable than the "abandonment of gold"? And now, as the result of an immense experience reaching through a period of over twenty years, I have arrived at the same conclusion with my friend Prof. H. S. Chase, and most emphatically endorse his enunciation, "In proportion as teeth NEED SAVING, gold is the *worst* material to use."

ARTICLE II.

AMALGAM.

"About the year 1826, M. Taveau, of Paris, advocated the use of what he called 'Silver paste,' for permanent fillings. Under this, as it were, shining title, was ushered into the world what was destined to be for years the Hydra of dentistry."—*History of Dental and Oral Science in America*, page 61.

THIS metallic preparation for filling cavities of decay in teeth was first brought to the notice of the dental profession in the United States, about sixty years ago, through the advertisements of two Frenchmen by the name of Crawcour.

It was called by them the "Royal *Mineral* Succedaneum,"—*succedaneum*, a replacer or substitute,—a name which is indicative of *fraud*, and which, consequently, stamps the adventurers as unworthy of professional respect.

Had these persons been unsuccessful pecuniarily, it is more than possible no notice would have been taken either of them or their filling material; but, as it was otherwise, and as they induced a really large number of the "best people" to submit to their operations, a shower of indignant epithets was heaped upon them by the incensed "first-class" practitioners of that day.

From the fact that, among other desirable qualities, the then new filling material was "easily manipulated," its possessors were enabled to fill a class of largely decayed teeth, with frail and broken cavity walls, such as had never been attempted by the most skilful operators then in practice.

This naturally reflected to their discredit and to the enhancing of the reputation of the Crawcours. It was in vain that asseverations as to the impropriety of retaining such "worthless" teeth in the mouth were made by those highest in authority. The comfort and satisfaction evinced by such as were enjoying the tangible benefits of the "new discovery" far more than counterbalanced the impression made by the scientific theorists.

This state of affairs, however, was not of long duration. Every occasional swelling of a face—attributable, as we now know, to existing conditions prior to filling, and which an equally tight stopping of any other material would have equally induced,—was accredited to the "Succedaneum," and *from this false basis* every sufferer was enlisted as an antagonist to the "Royal" filling material.

It was soon proven that, instead of the material being a *mineral* compound, it was purely metallic, and consisted of silver and copper rendered temporarily plastic by the addition of *mercury*.

This knowledge was eagerly spread abroad among the people; and every case of excessive flow of saliva—now recognized as a very frequent concomitant of periodontitis, and particularly of alveolar abscess—was pronounced *mercurial ptyalism;* and direful tales of wholesale loss of teeth and large portions of jaw-bone were freely circulated.

And yet, despite all this, the use of amalgam steadily increased. As years rolled by, the number of those practising dentistry, who employed it, was becoming seriously great. Gentlemen who had some claim to proficiency as manipulators and respectability as practitioners were admitting that its use was warrantable in some cases. Doubts were beginning to be expressed as to the truth of the objections which had been urged against it; and still more decided opinions were held as opposed to the *effects*—either local or systemic—which had been attributed to it.

Under these circumstances, it was deemed needful, for the proper maintenance of the dignity and purity of the Profession, that, as there had been organized an American Society of Dental Surgeons, such *official* action should be taken by that body as would place, beyond discussion, the line of demarcation between scientific regularity and *discreditable irregularity*.

As the initiative in this, a committee was appointed in the year 1841 for the purpose of reporting upon all filling materials of which *mercury* was a component.

This committee reported that the use of all such materials was injurious both to teeth and mouths, and that there was no tooth which could be *serviceably* filled that could not be filled with gold.

The report of the committee was adopted by the society *unanimously*.

But even this could not stay the onward course of the obnoxious compound; for, in the year 1843, it was found necessary to pronounce the use of amalgam "*malpractice*."

This declaration seems to have been the culmination of the *solid wave* of opposition; for, at this time, "information and facts" having been gathered and laid before the Medical Society of the County of Onondaga, New York, it was reported as their opinion that, although the "*mineral* paste" had *undoubtedly* produced *mercurial* (?) effects both severe and alarming, yet, nevertheless, the proportion of such cases was small when compared with the great number of instances in which it had been employed; "but that no care in the *combination* or *use* of the paste will prevent its occasional bad effects."

This was not the kind of report, by any means, that the opponents of amalgam desired, as it was calculated to weaken, rather than to strengthen, their cause, and did eventuate in more critical investigation of the subject at the hands of gentlemen who were not blinded by prejudice, but, on the contrary, were actuated by a desire to judge intelligently in regard to the matter.

Then it was that fillings of the decried material having done ten years of acceptable service in teeth that had been condemned by eminent practitioners as unworthy the effort to save, were brought under the observation of those who recognized evidences of its value in these enduring and continuing proofs of its capability.

Gradually some of the "better men" began to advocate its occasional employment, and so openly was this done, that the leaders of that day — 1845 — felt that the time for the most energetic action had arrived, and, actuated, as we believe, by a conscientious conviction as to the imminent danger which threatened the profession, and through it those who relied upon its members for the saving of their teeth, "*Resolved*, That a committee of investigation be appointed."

The duty of this committee was that of calling upon each and every member of the aforesaid American Society of Dental Surgeons and ascertaining his views in approval or disapproval of amalgam, and as to whether he used it in his practice.

Although this resolution was adopted, it was not until it had been most warmly discussed; and it is worthy of note, as illustrating the gradual changing of opinions, that Dr. E. Baker and Dr. Solyman Brown, both of whom were members of the committee of 1841, whose report has been referred to, spoke against the resolution.

The report of this committee was to the effect, that of nearly fifty members visited, less than a dozen ever used amalgam, and only three positively refused to pledge themselves not to do so. It further insisted that any amalgam was dangerous and unfit for use as a filling material, and concluded with this memorable sentence:

"*That any member of this society who shall hereafter refuse to sign a certificate pledging himself not to use any amalgam, and, moreover, protesting against its use, under any circumstances, in dental practice, shall be expelled from this society.*"

That any such action should have disgraced the records of any scientific body *in this country*, during this nineteenth century, is well-nigh incredible; but that it should have occurred within the ranks of dentistry, and that, too, while it was in the first throes of young life as a distinct profession, when liberality, brotherly love, and, above all, the most perfect freedom of scientific thought and investigation should have been tenderly nurtured, is simply monstrous.

It is comforting that we are able to relate the burst of condemnation which greeted this resolution of expulsion; hundreds of practitioners, society men, and others, while declaring themselves opposed to the use of amalgam, also declared themselves as much more opposed to the wretched policy which threatened punishment as the reward for investigation and experimentation.

At the next meeting of the society in 1847, for using amalgam and refusing to sign the required pledge, *eleven members were expelled.*

The society had now done its "worst," and in so doing, *it did its best!* Instead of interfering with the use of amalgam, even in the least degree, its action stimulated more practitioners to a critical investigation of the merits and demerits of a material which had so exercised the minds of the leading men as to have shaken dentistry to its centre.

The earnest workers wrought on till 1850; each year added to the proofs of the value of the condemned material; wretchedly compounded; wretchedly manipulated; wretchedly abused; it was yet standing, after a trial of nearly twenty years, in many a tooth which had been filled by the pretentious Frenchmen! More carefully compounded; with better manipulation; and judiciously used instead of badly abused, it had given such convincing evidence of its utility, and had even so long ago lived down so much of false aspersion, as that many of the members of the American Society felt it a duty that the action of 1845 should be revoked, and the error of judgment evinced in the proceeding be fairly acknowledged.

This feeling gave rise to more discussion, which resulted in the appointment of a committee to take into consideration the propriety of rescinding the former resolutions of "pledge" and "expulsion."

If the former action of the society was characterized by narrow-minded bigotry, the report of this committee was even worse. It is true, that it recommended that the resolutions which enforced the "subscription to the protest and pledge against the use of amalgam and mineral paste fillings for teeth, be, and the same are, hereby rescinded and repealed;" but it did so avowedly "*upon the belief that the resolutions had accomplished the object for which they were designed, and there no longer existed any necessity for their enforcement.*" Can this be credited? And yet this is the *record* of the transaction.

Again we are comforted in the knowledge that this contemptible method of treating so important a matter resulted in the immediate resignation of quite a number of the members.

Again the years rolled on; as truth was mighty and prevailed, so the society dwindled year by year, until at last, having lost all its former *prestige*, a meeting was held at which the President — Dr. Elisha Townsend — was directed to call a meeting for the consideration of "dissolution." At the meeting of 1855, a committee upon this subject was appointed. This committee reported against dissolution, was continued, and the meeting adjourned to meet in New York on the first Tuesday of August, 1856. At that meeting — *a very small one* — the following committee report was adopted:

" That we deem it expedient to dissolve this association, and that it be and is hereby adjourned *sine die.*"

ARTICLE III.

AMALGAM.—Continued.

IT seems to me, that, for the future good of dentistry, two facts are, at this time, worthy of careful, thoughtful consideration:—1st. That *amalgam*, which, from its misuse and abuse at the hands of those who attempted to utilize it, had become an object of unmitigated contempt to those who were recognized as representative men, has *practically* proven itself to possess qualities such as have *compelled* respectful recognition as a most valuable filling material; and,

2d. That such radical change of opinion in regard to *amalgam* had been wrought, *by some means*, during the passing of twenty years, as that the official acts which were esteemed needful in opposition to its use, *so far from accomplishing their purpose*, were directly instrumental in the final dissolution of the attacking organization.

These two points alone should furnish sufficient reason, to all who have professional interest in this matter, for the exercise of every effort at obtaining information in relation to it, and for the closest scrutiny of all statements, either favorable or unfavorable, regarding this very peculiar combination of metals.

But it is not alone the two facts of gradually growing recognition on the one hand, and final "dissolution" of organized antagonism on the other, that the career of *amalgam* presents for our reflection, but the equally significant ones that some of its strongest opponents were *early* found arrayed upon the side of its advocates, and were giving proof of the earnestness of their convictions by its frequent use in practice. Some of the men whose names are found upon the first records of the "amalgam war" as loudly outspoken in condemnation of the material, are those who in the later of the "twenty years" are but little less moderate in their views of its "unquestionable value in certain cases."

But, as though all this was not sufficient for *amalgam*—as though to show completely, and beyond all cavil, its inherent power for overcoming opposition—we find, "strangest of all," that in the very year—1855—when "dissolution" of the opposing organization was proposed, and when it was found that the American Society of Dental Surgeons actually did not possess sufficient strength to die (?), its President, Prof. Elisha Townsend, one of the best gold-workers of his day, gave dentistry *his sanction* to the first formula for the making of an alloy for amalgam that ever had the least pretension to "*respectability*."

This he did as the result of deliberate conclusions; as the result of long-continued, careful observations; as the result of an earnest desire to aid more than he had ever done—and that was *very* largely—in the preservation of teeth.

This he did *because* he saw, "daily, the evidence that teeth could be *saved* with amalgam which *he* could not save with gold."

As one of the most competent, the most enthusiastic, and the most conscientious practitioners of his profession, he possessed a *breadth of base* which permitted, nay, insisted upon, a mental recognition of merit wherever evidence of its existence could be shown.

As an ardent lover of his calling, he welcomed *any* aid in the development of its resources in the struggle for relief to suffering, and grandly proved his noble disregard of "self" by accepting that which he felt to be, at least in some degree, worthy, even though it came in the "questionable shape" of *old-fashioned, coin amalgam.*

He recognized its merits, ease of manipulation, capability for accurate adaptation to cavity walls, sufficiently resisting to subserve the purposes of mastication, and *proven* eminently tooth-preserving.

He recognized its demerit—*its one demerit*—discoloration: it turned black; it discolored the teeth; it disfigured the mouth.

The cry of "mercurial ptyalism" he had not found to be sustained by his observations. He felt that the material had been unjustly dealt with, unjustly maligned, unjustly condemned; and regarding it as one of the "defiant down-trodden," he resolved to give it aid and countenance for its merits, and to work, if possible, in the direction of its improvement.

In young life he had been taught the trade of the jeweller, and with the ideas thus inculcated, he naturally accepted, as an improvement, the "refining" of the metals.

It was thought that by the *purifying* of the silver, the elimination of the copper, — which was then held to be a very objectionable ingredient, — and the addition of a large proportion of *pure* tin, an alloy could be produced which would be decidedly less objectionable than the silver coin.

The proportions of this alloy — afterward known as Townsend's — were

    Pure Silver . . . . 4 parts.
    Pure Tin . . . . 5 parts.

The metals were thoroughly mixed while molten, and were then cast into matrices so shaped as to form ingots suited either for filing with a coarse file, or for turning into shavings in a lathe: these shavings were then rubbed into a sort of powder.

To the required quantity of alloy, thus prepared, was added a portion of mercury, and these were mixed in the palm of the hand by kneading with a finger. By this means a metallic mixture of doughy consistence was obtained, which might very appropriately be called a "silver paste."

This was then further "purified" by a process known as "Washing." For this, the amalgam mass was placed in a small porcelain mortar and rubbed with a pestle, after having had added to it a little absolute alcohol; the result of this rubbing was a decided *blackening* of the alcohol, which was then poured off, and, with the addition of more alcohol, the work of rubbing was repeated; after two or three such "washings" — as they were styled — much less "blackening" was produced, and it was therefore assumed that this process would prevent, or at least diminish, discoloration. The mass was then placed in chamois skin, and the surplus mercury having been "squeezed" out by pliers, it was packed, artistically, into cavities of decay.

As we look at all this, with the aid of such light as has been thrown upon the subject by years of systematic, scientific, metallurgic experimentation, it seems very like many other lines of work prosecuted with the very best intentions, but based entirely upon the crudest and most fallacious assumptions.

Not one of the various proposed "purifying" processes was possessed of any value; and, what is indeed most peculiar, *not one is other than detrimental!*

A filling material was made which is *not nearly so tooth preserving*, and which, though it is prevented from discoloring — in limited degree — by the addition of tin, is, by this modification, caused to "shrink" most notably, and from which is taken, practically, all its "edge-strength." The "washing" process, so far from being advantageous, is proven to be otherwise, and, at the present day, nearly all manufacturers of alloy for amalgam *caution against it* in italics or capital letters!

And yet, the dental profession, in the persons of its representative men, welcomed with open arms the *tinned* and *"washed'* amalgam, as introduced by the President of its American Society of Dental Surgeons, and fairly revelled in the seemingly inexhaustible capabilities of this "purified" sinner!

The largest and most inaccessible cavities in the frailest and "most worthless" of teeth were filled just as the Crawcours had filled them! Crowns were built upon the roots of molars and bicuspids, and were viewed with complacent pride, with no thought of disgrace, with no fear of mercurial ptyalism!

Some enthusiasts even went so far as to make rows of crowns upon cuspid and incisor roots, and then stepped back to admire.

This was a change, indeed. It is true, there were *some* who never joined in this acceptance; but it is also noteworthy that the later record of such has not placed them in any position which entitles them to rank as equal in professional attainments with most of those who warmly espoused amalgam.

This state of things continued but for a few years, — some three or four, —for, in this length of time, not only had "unworthy" men, in numbers, abused the trust offered under the mantle of respectability, but much crevicing between fillings and teeth had taken place from undue shrinkage, and much breaking away of edges, both of fillings and cavity walls, was only too apparent; but, worst of all, the "purification" *by fire and by alcohol* had not proven equal to the emergency; and though the discoloration of fillings and of teeth was not nearly so great as with the coin amalgam, yet it was, nevertheless, far too much for the properly æsthetic ideas of good gold-workers, and thus the tide began to turn.

Again, it was *now time* for much of the pathological sequellæ which naturally pertains to teeth that, having been largely decayed and filled, permit a more or less gradual, and more or less painless devitalization and disintegration of their pulps. And thus it was that with quite an amount of disappointment in regard to the frequent occurrence of plug-discoloration, and with increasingly unsatisfactory statistics in connection with subsequent peridental irritation and alveolar abscess, the dental mind became exercised in such degree that some of those who were then recognized as leading men obtained from Prof. Townsend the following "recantation," as it was called, which was published in the "Dental News Letter" of April, 1858.

"For the *Dental News Letter*.

"AMALGAM.

"MESSRS. EDITORS.—I promised to report to you any change in my practice in the use of amalgam for filling teeth, founded upon further experience. In all that I have ever said or written upon amalgam, I have been very careful not to advocate its use except in those cases which could not be filled with gold, and where extraction was the only alternative. I find my name has been used as authority for its indiscriminate and unlimited use, which I certainly never intended or supposed could happen.

"I wish now to say to the profession that I have entirely abandoned it, and shall never use it again in my practice. I have come to this resolution for reasons which I will now state. In many of the cases where I most relied upon it, and expected to have the best results, it has entirely failed; as in the buccal cavities in molars, when they extended beneath the free margin of the gum, I found that while in some mouths the material remained white and clean, in others it became very black in a few days, and in almost all cases, upon removing the filling, the under side was blackened, and the same color given to the tooth. Again, in the infirm teeth, for which it seemed the only thing, and for which it was best adapted by its plastic nature, many of them had to be removed, owing to suppuration of the gums, caused by the tight closing of the previous vent for the escape of pus.

"Therefore I have come to this broad conclusion, that a tooth so infirm as to need a soft filling would be best removed, for the health of the mouth and the health of the patient; and that my practice hereafter will be to advise their removal, and then leave the responsibility with the patient. ELISHA TOWNSEND.

"No. 1606 Locust Street."

As it was entirely owing to Prof. Townsend's faith in the tooth-saving quality of amalgam that I had been induced to commence experimenting with it, and as I had gradually come to regard even the discoloration of coin amalgam as a matter of comparatively little moment in non-conspicuous places, when contrasted with the comfortable and, as it seemed, durable saving of a large class of truly valuable teeth, it was with peculiarly painful professional feelings that I viewed this retrograde step — as I thought it — in that work which I esteemed a *progressive* movement.

As it was, the *latter part* of the "recantation" took from it all its power for controlling me.

I had *just then finished* a line of work upon the *systematic treatment* of such teeth as were specifically classed as appropriate for amalgam, with the view to preventing the occurrence of such untoward results, and was then instructing my private pupils in such pathology and therapeutics of alveolar abscess as I have since taught and demonstrated in my lectures and clinics.

I had also entered upon that protracted individual labor which culminated at the meeting of the American Dental Association of 1875, when I joined my friend, Dr. S. B. PALMER, with the conviction that *his theory* regarding the need for "compatibility" between filling-material and tooth-bone, as an essential for success in saving teeth, *satisfactorily explained* the results which my accumulated statistics of twenty years had placed before me.

I engaged in that work with the full recognition of three basal positions in connection with amalgam:

First. That one of the very best operators of his time —

Prof. Townsend — had conceded that *amalgam* was a material possessing, from some cause, a power for saving teeth which exceeded that of gold, even when worked with his remarkable manipulative skill.

Second. That all previous opposition to, and aspersion of, *amalgam* had been founded upon *complete ignorance* of it; while the evil results which were ascribed to it were certainly attributable to other well-recognized causes; and,

Third. That the efforts at correcting its demerit, though well-meant, were misdirected.

It is at this point that I wish it to be distinctly understood and remembered, that the *positively deteriorating* effects of the change of formula from silver, copper, and mercury, to tin, silver, and mercury, as accepted and advocated by Prof. Townsend, were not recognized for many years. It was not suspected that the "shrinkage" of the so-called Townsend's amalgam exceeded enormously that of the silver coin amalgam; indeed, it was believed and taught by Prof. Townsend that amalgam *expanded* during its crystallizing, hardening just as does water while passing from the fluid to the solid condition in freezing.

It was not recognized that the sustaining "edge-strength" of the old amalgam was seriously impaired by the lavish incorporation of tin. It was not known that the increased "bulging," due to the less controlled mercurial tendency to assume spheroidal shape, was the cause of "crevicing," such as was never seen in connection with the old material.

It was not known that the "setting" of the new amalgam was even slower than that of the old; for few, indeed, of these "respectable" experimenters knew anything of the setting of any other compound.

It was alone recognized that the discoloration, though modified, was not sufficiently so to meet with anything but general disapproval. This was the situation when I commenced amalgam work in 1855.

The first positions to *establish* were those of the truth of the assumption that amalgam possessed a "tooth-saving power" which did not pertain to gold; that instead of being "dangerous and unfit for use," it was advantageous, and, under certain circumstances, eminently fit for use; and that instead of being "malpractice" at all times to employ it, it was decidedly "malpractice," at times, not to bestow upon patients the great good of its utilization.

This could only be done by first learning its proper mode of manipulation; then filling a large number of cavities in teeth, where its behavior could be compared with that of gold, and afterward waiting for a sufficient number of years in order to determine this. Accordingly, I resolved to fill some hundreds of cavities with amalgam in teeth such as were usually extracted, or were only filled either by protracted operations with tin-foil, or by equally tedious and vastly more expensive operations with gold.

As this work progressed,—done, as it naturally would be, with the only amalgam with which I was practically conversant, "Townsend's,"—the possibilities of the material presented themselves with increasing clearness; and thus it was that, led on step by step, I gradually *treated* and filled *comfortably* a class of teeth which, even to my developing conceptions, was something extraordinary, until, within two years, it seemed to have become the task to question, not the necessity for extraction, but the limit to possibility for salvation.

I had decided upon *five years* as a reasonable length of time for the determination of the question of "superiority as a tooth-saving material," and so markedly had this been shown at the expiration of that period, that I should certainly have questioned my own ability with gold were it not that *in many mouths* I had also had the opportunity of contrasting with amalgam work many fillings of gold, of comparatively recent introduction, which were the work of those who were deservedly ranked as "first-class" operators.

At the close of the *five years* of probation, the time did not seem so long to look back upon as it had seemed to look forward to, and so I gave another year.

At the end of this time, I, too, became convinced that *I* could "save teeth with amalgam which *I* could not save with gold," and that *therefore* it was *my duty* to myself, and to those who were committed to my charge, to use the means, the value of which I had so thoroughly tested.

ARTICLE IV.

AMALGAM.—Continued.

FROM 1861 to 1875, a period of fourteen years, my progress in the substitution of amalgam for gold had been very considerable, and had been attended with very satisfactory results; this, however, had been largely due to the "combination" fillings — oxy-chloride of zinc, sub-strata, and linings, which I had used since 1860; but it was also due, in part, to my gradual change of formula from that of the so-called "Townsend's."

This formula, "Townsend's," had also gradually changed; but while it had done so by greater addition of tin, and had become 6 parts of tin to 4 of silver, the alloy to which I had been led was composed of 9 parts of tin to 13 of silver.

I had noticed the wonderful maintenance of integrity on the part of the silver coin amalgam fillings; and though I then shared the prevailing prejudice against the copper which I knew such amalgam to contain, I was, nevertheless, compelled to admit the loss of edge-strength, the greater crevicing, and the general inferiority of amalgam made from the largely tin alloy. For this reason I commenced the increase of silver, and, notwithstanding the loss in the direction of "maintenance of color," continued its addition until I had reached a decided preponderance of that metal.

The superficial discoloration had long since lost much of its power to stamp the filling as objectionable, in view of extended and long-continued proofs of capability for tooth salvation; and even the objection of tooth discoloration had been thus early largely overcome by "lining."

So much work seemed to offer in other lines, and so urgently did it press its claims, that nothing like *sufficient* attention was bestowed upon alloys for amalgam until the papers upon that subject which were presented at the meeting of the New York Odontological Society, held in December, 1874, came to my notice. These I read with the greatest interest and with the most critical circumspection, and I soon found that I must either reconsider the whole of my years of experimentation, or take decided issue with the evident tendency of all manufacturers of alloys.

But, again, I felt that I must receive even the statements contained in those papers with doubting hesitation, for, with the exception of the experiments and conclusions of Dr. S. P. Cutler, of Memphis, Tennessee,—with which, in the main, I could agree,—there was a confused mass of assertions, experiments, analyses, and deductions, which, while they contained much that seemed reasonable, and stated some facts with which I was conversant, yet also contained so much that was conflicting or of doubtful value, or was, in my opinion, absolutely incorrect, that I judged it to be for the most part a questionable contribution to knowledge.

I found reference made to the practical experiments of Messrs. John and Charles Tomes, 1861 to 1872, which I had regarded as conclusive, and had accepted as such. I found the term "oxidation," as almost invariably applied to the discoloration of amalgam fillings, corrected by the acceptance of the *long before* urged, and much more reasonable, hypothesis of "sulphuretting." I found confusing statements, such as an alloy of "two parts silver, one part tin, and about twenty-five per cent. gold," which would be simply, two silver, one tin, and one gold. I found a long list of elaborate experimentation with filled teeth and amalgam pellets, weighed with marvellous accuracy, placed in little bottles containing saliva acidulated with nitric, acetic, citric, or hydrochloric acid, and kept in a water-bath inside of another water-bath, at a uniform temperature — blood-heat — for a period of three months, in order to prove by analysis of the saliva whether or not amalgam fillings would be capable of producing mercurial ptyalism.

I could not reasonably doubt the certificate of the "Professor of Analytical and *Applied* Chemistry," that the saliva "contained no mercury in solution;" but I gravely doubted the value of the test as proving anything whatever in relation to the systemic influence of amalgam fillings.

I found analyses, made "through kindness," which, although so full of such glaring discrepancies as scarcely to deceive one who had closely observed in amalgam work, were yet so speciously presented as to grossly mislead the mass of inquirers, and thus prove anything but a "kindness" to them.

In these the information was given that "Lawrence's" alloy had in it nearly fifteen per cent. of copper! and "Townsend's Improved," which sold for $2.00 per ounce, was given as composed of materials—silver, tin, and gold—in such quantities as would have cost $1.80 per ounce for these alone!

From a personal acquaintance with the dealers in alloys, I thought *this* could not be true.

About this time I came into *metallurgic* association with my friend, Mr. Patterson Du Bois, one of the assistant assayers of the Philadelphia Mint; he had been for some years a patient of mine, and recognizing the deep interest which I took in dental alloys, he offered to make for me a series of analyses.

I eagerly embraced this opportune aid to investigation, and the work was commenced. So great an interest upon his part was soon engendered, that he asked permission to add the power of another assayer to the task; thus Mr. Jacob B. Eckfeldt became interested with us, and we prepared for extensive and thorough manipulation. The incentive then, to them, was the probability of eventually arriving at an excellent alloy which they could manufacture. The incentive to me, was the evident possibility of so directing experimentation as to secure, promptly, a good result, based upon some reliable data.

This I felt would be a boon to my profession, my patients, and myself.

Our first requirement was the purchasing and analyzing of one or more samples of every alloy of note in the market. It was needful that caution should here be used lest attempts should be made to lead us astray. It was recognized that manufacturers might, very properly, regard it as "business" to prepare for us *special samples*, if it was known that we purchased expressly for analysis; we, therefore, as promptly as possible, and largely through the hands of other parties, obtained the desired samples from various cities and countries; of the most noted we obtained three samples each at various times, ranging from one to three months apart.

As one after another of these alloys were analyzed, two salient points became apparent; *first*, that the analyses given in the paper compiled by Prof. T. B. Hitchcock, and presented by Prof. T. H. Chandler, at the New York "Odontological" meeting of December, 1874, were almost entirely unreliable, and eminently calculated to mislead inquirers; and, *second*, that with but two exceptions, viz., the alloys of Drs. Lawrence and Hardman, all the other alloys in the market were based upon one of two ideas, viz., some regarded the addition of *small quantities* of gold and platinum as essentially advantageous, and others did not; but practically all agreed that the proportions of tin and silver should be from fifty-five to sixty-five per cent. of tin, and from thirty-five to forty-five per cent. of silver.

As examples in proof of these positions I offer the following half-dozen analyses:

ANALYZED BY DR. E. S. WOOD, PROFESSOR OF CHEMISTRY IN HARVARD MEDICAL AND DENTAL SCHOOLS.

"*Townsend's.*"
Silver 40.21
Tin 47.54
Copper 10.65
Gold 1.06

"*Arrington's.*"
Silver 40.00
Tin 60.00

"*Walker's.*"
Silver 34.89
Tin 60.01
Gold 4.14
Platinum 96

"*Townsend's Improved.*"
Silver 39.00
Tin 55.69
Gold 5.31

"*Lawrence's.*"
Silver 47.87
Tin 33.68
Copper 14.91
Gold 3.54

ANALYSES BY MESSRS. ECKFELDT AND DU BOIS, METALLURGIC SECTION OF THE "NEW DEPARTURE CORPS."

"*Townsend's.*"
Silver 42.00
Tin 58.00
Copper, none.
Gold, none.

"*Arrington's.*"
Silver 42.50
Tin 57.50

"*Walker's.*"
Silver 30.50
Tin 69.00
Platinum 50
Gold, none.

"*Townsend's Improved.*"
Silver 43.00
Tin 57.00
Gold, none.

"*Lawrence's.*"
Silver 47.50
Tin 47.50
Copper 5.00
Gold, a trace ($\frac{1}{100}$th gr.)

"Johnson & Lund."		"Johnson & Lund."	
Silver	38.27	Silver	38.50
Tin	59.58	Tin	59.40
Platinum	1.34	Platinum	.40
Gold	.81	Gold	.60
Cadmium not mentioned.		Cadmium	1.06

As I had experimented in the mouth with nearly every alloy analyzed, and was thus conversant with the working qualities, the defects, and the advantages of each, it required but a few weeks to make the necessary deductions for the initial alloy. Regarding the average proportions of tin and silver, in nearly all the alloys manufactured, as 60 tin and 40 silver, and informing my co-laborers of the advantages which I had noted in the increase of silver, we, after nine *preliminary* experiments, *inverted* the proportions, and *commenced* with 60 silver and 40 tin; and it is with much satisfaction that I can say, that five years of experiment and observation, done, as we believe, with *unbiased* care and accuracy, has only strengthened our convictions that these *basal* proportions — modified by *additions* of copper, gold, and zinc; possibly antimony; *more doubtfully*, platinum; and *most doubtfully*, cadmium — will constitute the best alloys for amalgam fillings which can be made *from the metals that are now accepted as components for such alloys.*

At this point, I desire to enter a protest against the tendency to mislead, which is found in names and statements pertaining to alloys, as I regard the tacit acceptance of them by dentists as a *reproach to dentistry.*

Every "gold and platina" alloy in the market is, *at present,* composed of more than fifty *per cent.* of tin, and more than forty *per cent.* of silver, while the remaining few *per cent.*—two to seven, only—are found to be of gold and platinum. Now, as the addition of a small proportion of gold *in connection with so large a proportion of tin,* even though it be *in slight degree* advantageous; and as the addition of platinum, either in small or large proportions, is, *as yet,* not proven to be of any benefit, but must be ranked as of "doubtful" utility, these so-called "gold and platina" alloys cannot be, in much degree, superior to the ordinary alloys of tin and silver. I therefore submit the proposition, that, both in *name* and *inference of great superior-*

ity, these alloys are deceptive, and their acceptance is derogatory to the intelligence of dentistry.

Many alloys are constantly advertised with the assertions of "no shrinkage" and "no discoloration," when the *truth* is that fillings made from them both "shrink" and "discolor" most notably; and those that shrink least, as a rule, with very rare exceptions, discolor most; while those that discolor least, as a rule, shrink most! *Alloys are constantly appearing which are advertised as "superior," as "best," as "made by a new process," etc., which, upon analysis, are found to be composed of the same ingredients, in the same proportions, as have been almost universally used; their "plasticity," their "setting," their "edge-strength," their "shrinkage," their "color-tests," are all *exactly the same* with the other *average* alloys, and yet each has a long list of "testimonials" as to its *superior peculiarities* from gentlemen who are regarded as eminent in our profession.

NOTE.—During the session of the American Dental Association at Niagara Falls, in 1878—only *ten* years since—I was told by one of the prominent delegates that he preferred *decidedly* the "Arrington" alloy to that which was sold as "Townsend's;" and upon asking him "Why?" the reply was, "because it works much more plastic, sets much quicker, makes better edges, keeps its color better, and altogether makes a *much better filling*." I could hardly credit my own hearing! for *I knew*, from analyses, from working, from careful testing, and from many observations, in the mouth, that the two alloys, as compounded and cut, were *precisely the same*, and that the only difference between them was that of *size of grain*.

It is a satisfaction to know that this kind of thing *must pass away* in due time. Already the advanced collegiate instruction provides students with the needful information, both theoretic and *practical*, in the direction of testing and analyzing alloys which are offered for sale, and also with instruction in the compounding and making of all such alloys and other materials for plastic filling as have been proven, by competent testing, worthy of confidence.

Already the *young men* go into the field of practice so much better "armed and equipped," in relation to these matters of vital interest and importance, that they are self-dependent, and can far better instruct than be instructed by those whose experience in plastics has been guided by "judgment" rather than by knowledge.

* These remarks refer to the *usual* three or four metal alloys.

ARTICLE V.

ATTRIBUTES OF METALS USED FOR AMALGAM ALLOYS.

TWO or more metals combined by melting form an "alloy." One or more metals held in combination with mercury, *by the mercury*, form an "amalgam."

This is the dental acceptation of these terms; and we therefore find "alloys," containing mercury in very small proportion, added for the purpose of making them more fusible; but no mixture of metals is regarded as an "amalgam" unless it has mercury as one of its components in sufficient quantity to exert a *combining influence* over the other metals.

Our present theory regarding the formation of amalgam is, that metals which melt at comparatively high temperatures, being so prepared as to permit of the fusing influence of mercury,— a metal which is "molten" even at very low temperature ($-30°$),— are, by admixture with mercury, melted into union with it.—Du Bois.

The acceptance of this theory not only seems to simplify the otherwise complex and mysterious behavior of this peculiar filling-material, but aids much in the accurate construction of various alloys with the view to accomplish specific purposes.

Thus, the "setting" of the amalgam — as the gradual *hardening* of the mass is technically termed — is recognized as due to the *secondary* cooling influence upon the mercury from mixture of such metals as are not molten at ordinary temperature; and it is readily appreciated,— at least, in a general way,— that the higher the melting-point of the admixed metal, the quicker will be the "setting."

I say, "in a general way," for it is practically demonstrated — by platinum — that, in some way,— from physical characteristics,— a metal even of *very high* melting-point may not produce this result; but this is *exceptional*.

The decided "bulging"— as the *spheroiding* of the mass is termed — of fillings made of alloys largely composed of tin, may be plausibly attributed to the combined ductility and low melting-point of the admixed metal. Under the strong con trol of a rapidly and rigidly hardening metal, this spheroiding is found to be limited; under the feeble control of a metal which cools slowly, and when cold is yielding, the gradual spheroiding continues for many months, and even years, until it eventuates in excessive "bulging."

The "shrinking" of the mass is also in general, though not in universal accord with the contraction of the metals composing it, as they pass from the fluid to the solid state. Thus, tin, losing its fluidity slowly, permits shrinkage in proportion as it is added in quantity. Gold and copper, which lose fluidity quickly, and silver, which *expands* in cooling, notably lessen shrinkage. Platinum — by far the most obdurate of metals used in dental alloys — seems to exercise such imperceptible control over shrinkage as to be unworthy of mention in this connection.

The "plasticity" of an amalgam, using the term as applied to the length of time during which it may be worked, as "it remains plastic for fifteen or twenty minutes," is almost entirely in consonance with the relative proportion of the easily melted tin; but here again we find platinum *increasing* this peculiarity, and even in some alloys maintaining a plastic condition in a most undesirable degree.

The "edge-strength" of amalgam is singularly influenced by the ratio of edge-strength of its components, *as compared with their melting-points*; thus, some metals which have little edge-strength and melt at low temperatures, as tin, require to be mixed in large proportion in an alloy before they will seriously impair the edge-strength of an amalgam; while some metals which have fair edge-strength and require a high temperature for their melting, as copper, in comparatively small proportion of admixture, will *markedly* impair the edge-strength of amalgam; metals which have reasonably good edge-strength and melt at high temperatures, as silver, add reasonably to the edge-strength of amalgam as their relative proportion in an alloy is increased reasonably; other metals of fair edge-strength which melt at high temperature, as gold, add notably to the edge-strength of amalgam, especially that made from largely silver alloys, even though added in but small quantity to the alloy.

In effect upon edge-strength platinum seems to be entirely

neutral, whether added in small or large quantity. This was tested in amalgams made from alloys containing platinum in varying quantity, from less than one per cent. to more than five per cent.

The "hardness" or "density" of amalgam *seems* to depend entirely upon the *toughness* of the metals composing the alloy. This is a point which would require a vast amount of experimentation to establish; an amount which, at present, would be non-compensating, for the reason that we have no difficulty in making amalgam of sufficient density. Indeed, if we could improve in the direction of "maintenance of color" without increase of "shrinkage," and without loss of "edge-strength," but by diminishing in "density," it would be rather beneficial than detrimental, for even the least desirable of all the alloys, Townsend's, Walker's, Arrington's, etc., are more than sufficiently "dense" to subserve all practical purposes.

NOTE.— The terms "hard" and "dense," though not synonymous in metallurgy, seem to have become such in the technical phraseology of amalgam alloy work; they are used as opposite to "soft," while the term "brittle" is employed as indicative of another characteristic which may pertain either to soft or hard amalgam, it meaning *easy to be broken or crushed*.

The metals which are at present used in alloys for dental amalgams, are, silver, tin, copper, gold, zinc, antimony, cadmium, and platinum.

NOTE.— Palladium, tellurium, etc., are not included in this list, because the several reasons, that the very high price of the metals — from thirty-five to fifty dollars per ounce — the great difficulty of obtaining them in adequate quantity and the decidedly *theoretic* value yet pertaining to them as components of amalgam alloys, were deemed sufficient to deprive work upon them of all practical value, at least so far as pertains to present demands.

SILVER. *Argentum.* Symbol, AG. Fuses at 1873° F. *Expands on cooling.*

As in 1826, under the auspices of M. Taveau, the name of "silver paste" was given to the prototype of that filling material now known as "amalgam," so might it reasonably be given to all the amalgams of the present day.

Silver is the first, the most important, the essential metal of a good amalgam alloy for filling teeth; it is the largest component of every truly good "submarine," "usual," or "contour" alloy in the market; and it will be found by all actual tests, that just in proportion as an alloy contains silver *in small amount*, so will it make an amalgam which will *not compare favorably*, in its meeting of *almost* every dental requirement, with those made from alloys which are largely composed of this metal.

The most ordinary amalgams of to-day, taking that made from "Townsend's" alloy, 60 tin, 40 silver, as the type, are mixed into *pastes*, with about two-fifths their weight, consisting of mercury, a little less than two-fifths tin, and a little more than one-fifth silver, 38 parts mercury, 39 parts tin, 23 parts silver = 100 parts; and yet the silver is the *essential*, for without it the amalgam would be worthless for filling teeth. In the next grade of alloys are those composed of the same relative proportions of tin and silver, but with the additions of gold or gold and platinum.

In these the silver is still the *essential;* but in the endeavor to keep its proportion small, and to overcome, by the addition of the other metals, the loss of edge-strength and the tendency to shrink and bulge incident to large admixture of tin, it is found that *something* is yet required to meet the emergency; and it is therefore proposed, by the manufacturers of such alloys, that they be mixed with the smallest possible amount of mercury, be worked in the form of a "kind of powder," and be made into amalgam by packing into the cavity with warm instruments.

With the next grade of alloys the stereotyped proportions of tin and silver are inverted, of these the "old reliable" alloys of Hardman of Muscatine, Iowa, and Lawrence of Lowell, Mass., are the types; but in these we find the peculiar admixture of copper in addition to the increased proportion of silver.

Any operator who has, for many years, wrought faithfully with these alloys, has found no need for "dry powder" in place of well constituted plastic "amalgam," nor has he had to heat his instruments in order to introduce his fillings; and I am sure that each can point to many a monument of "capability for tooth salvation," even in desperate cases, which is alike a marvel to himself and to his patients; and these results, both that of possibility of easy working, and that of eminence in tooth-saving, are due, almost exclusively, to the large proportion, relatively, of *silver*.

Again, with progressing gradation, we reach the present highest type of "usual" alloys, composed of silver, tin, gold, and copper, and represented by the very properly named "Standard" alloy of Eckfeldt and Du Bois.

Also, that excellent type, than which no other has so well subserved my purpose for nine-tenths of all the amalgam work I have done during the past six years, the "Contour" alloy of 64 silver, 33 tin, 3 gold.

And again, that peculiar "Front Tooth" combination of 50 silver, 24 tin, 20 gold, 6 zinc, which, while it has, in turn, been practically set aside by the heavily coppered alloys (from 20 to 40 copper), even yet enters as a component into the best of the "Front Tooth" alloys.

In all of these we find the relative importance of the silver very fully recognized; and while we cannot regard them as superior to, if even equal with, the two metal alloys, when viewed merely as tooth preservers, we yet cannot deny to them the possession of sharper and stronger edges; greater rapidity of setting; finer plasticity in working; and a better general maintenance of color.

It will be noted that I give to silver a position of decided pre-eminence; and this is done with the view of impressing my conviction that of all the metals which have ever been used for the saving of teeth, silver is the one which is entitled to the greatest praise, the greatest confidence, and, therefore, to the greatest consideration.

To the *silver*, then, we ascribe the position of metal of first importance; it is essential to the proper setting of the amalgam; it notably maintains the bulk integrity of the filling, and it forms, with the sulphuretted hydrogen, which is more or less constantly brought into contact with it in the mouth, sulphide of silver, which, though it discolors the filling and also the tooth, is nevertheless highly conducive to the permanent saving of teeth which are not only largely decayed, but markedly predisposed to continued decay.

TIN. *Stannum.* Symbol, SN. Fuses at 442° F.

I have placed this metal *second* on the list, because, although it is a component of comparatively modern introduction, and is largely ante-dated by copper, and although its introduction *well-nigh ruined* a material which had bravely withstood the assaults of unscrupulous prejudice for more than a quarter of a century, it has at length, by long and elaborate experiments, been assigned a position in which, under control, it becomes the *second* in importance upon the grounds of *quantity* and *usefulness*.

All such alloys as I should favorably regard, have from 20 to 55 per cent. of tin; it is found that by the addition of copper and gold, both antagonists of "shrinkage," the most deleterious of the effects of tin can be counterbalanced; that under this control sufficient silver can be used to obviate a detrimental loss of edge-strength; that the retardation of "setting" is prevented, and that the tin not only loses its power for harm, but becomes an ingredient of manifold utility; it greatly augments the facility of amalgamation; it aids in producing a good color and in preventing discoloration; it diminishes conductivity, and is our best flux in the making of alloys.

COPPER. *Cuprum*. Symbol, Cu. Fuses at 1996° F.

The "Royal Mineral Succedaneum" was an amalgam of mercury, silver, and copper; it was made from silver-coin filings which were composed of an approximate to nine-tenths silver and one-tenth copper; the amalgam was made by mixing about equal parts of mercury and filings; the result was a mixture consisting of 50 parts mercury, 45 parts silver, 5 parts copper = 100 parts.

This material saved teeth even though abused and misused, but its surface discolored outrageously,—it darkened teeth abominably,—and the soluble salts of the uncontrolled copper in fillings entering pulp cavities and canals, so permeated the dentine and cementum, coloring them a livid green, as to give rise to a warrantable suspicion that to the presence of this metal might be attributed the subsequent death and exfoliation of such teeth.

NOTE.— As I cannot but admire the ingenious reasoning of Mr. Tomes in regard to this form of discoloration, I quote from him:

"The sulphide of copper, under the influence of exposure to air and moisture, readily becomes oxidized, and forms the sulphate. Hence it is almost certain we shall have sulphate of copper formed upon the exposed surface of the filling. Now this sulphate is freely soluble, and hence is likely to permeate the dentine, when it will again be converted into sulphide, whilst the sulphides of other metals, not being so readily converted into soluble salts, will not so thoroughly permeate the teeth."

For this reason copper was early looked upon as a very objectionable ingredient of amalgam, and "virgin silver" was the first *desideratum;* but I have constantly noticed, during all the many years in which my professional brethren in our own country have been reaping the harvest of a fearful loss of soft teeth, under the *regime* of elegant gold-work, and have been vainly endeavoring to stay the tide of destruction by an increase of the already lavish incorporation of tin in amalgam, that my professional brethren in other countries, particularly England, have been saving just such teeth by using amalgam, of which copper was a component; and that in some of these filling materials this metal abounded in marvellous quantity; thus it was that, as I was continually *lessening the tin*, and increasing the silver, in my efforts to save the teeth which had become wrecks from being repeatedly filled with gold, I began thinking favorably of the despised and discarded copper.

I was not alone in this, for others seemed so evidently impressed with the value of the diminishing—silver—and abandoned—copper—metals, that the heavily silvered *three-metal* alloys were produced. It is one of the peculiarities of the "judgment" practice, that hundreds of practitioners have, for a long time, used Lawrence's and Hardman's alloys with most satisfactory results to themselves and to their patients, who would have been *horror-stricken* if their "knowledge" had told them that each contained copper!

NOTE.—It is but little more than *ten years* since a paper upon the subject of "amalgams" was read by a *Professor of Dental Metallurgy*, from which I quote: "Most dentists would hesitate about employing an amalgam known to contain copper; yet I was *much surprised recently* (!) at the result of a careful analysis of an amalgam filling which had done good service for twenty-three years, during which period it had retained a perfectly bright and smooth surface exteriorly. . . . I found its composition — after freeing from mercury — to be tin, 55 parts; silver, 40 parts; copper, 5 parts."

In view of this statement, and in view of the extended espousal of "pure" copper amalgam at the present time, it would seem that some of the foundations of dental practice *were* not, and possibly *are* not, as *solid* as might be desirable!

But it is not alone the marked compatibility of these partially copper amalgams with tooth-bone which characterizes them as peculiar; for it has also been noted, and commented on, that pulps seem to evince decided toleration for them. It has been my own experience in practice, that teeth, in which "deep" and "very deep" cavities of decay had been filled with amalgams containing copper, have, as a rule, behaved better than those which had been filled with amalgams which did not contain this metal. My record of pulp-saving—noted in each case after *five years* of trial—is very significant in its connection with the presence of a portion of copper; for, while I do not presume to say that it is, as yet, proven beyond question that pulps do maintain their vitality better under amalgam fillings which contain copper than under those which do not, I must nevertheless admit that observation has seemed to point so conclusively in this direction, that I should, in my practice, permit, with the exception of tin, no metal filling which did not contain copper to approach a pulp.

And yet another direction in which copper seems useful as an adjunct, is its singular properties as alloyed with tin — the alloys of tin and copper — "hard brass;" "bronze" may properly be regarded as among the most curious of metallic mixtures. From these soft and yielding metals are compounded some of the best alloys for brass ordnance; the unyielding "anti-frictions," and, even more peculiar still, the obdurate "speculum metal."

Very *white* alloys result from admixture of copper with tin, and this effect is noted in amalgam; by proper manipulation—which will be given in place — the face of such an amalgam filling can be finished with exceeding *whiteness;* and although I cannot but think that in certain mouths the presence of the copper tends much toward increasing discoloration, yet, the *usual result,* particularly under the controlling influence of an addition of gold, is the making of very light-colored fillings, which retain their color remarkably well.

To copper, then, is ascribed the power of diminishing, expansion, and "shrinkage," as copper amalgam *contracts much less than tin;* it favors rapidity of "setting;" it is accredited with power to add to the "compatibility" of the filling material with tooth-bone, and thus better saves the teeth; it is thought to produce greater harmony between filling material and dental pulp, and thus promotes "toleration" of foreign material, in close proximity, on the part of that organ; it adds to the immediate whiteness of the amalgam, while its tendency to gradual discoloration is notably under control; for these reasons I regard it as possibly a valuable component of dental amalgams.

GOLD. *Aurum.* Symbol, AU. Fuses at 2016° F.

This metal is one of those more recently added to the list as an ingredient of dental amalgam alloys; it has been but very few years—eight or ten*—since systematic experiments with it were inaugurated; opinions are yet varied in regard to some of its effects upon amalgam, but enough are in unison to show that it is worthy of place in our considérations, while the concluding work of the "New Departure Corps" left it with me. so far as its *possibilities* are concerned, in the position of the most *undetermined element.*

In relation to its effect upon "shrinkage," there can be but one opinion; the "specific gravity" work of Dr. C. S. Tomes established not only the fact of its prevention of shrinkage, but the ratio of its preventive power; but opinions concerning its control of color, and particularly its control of "setting" and "density," are antagonistic in the extreme.

In the paper of Prof. Hitchcock (Harvard Dental Department), it is stated that an English alloy of *gold* one *part*, tin two parts, silver three parts, *is said* to keep its color well, but does "*not become very hard.*" And again, he says, "Gold does *not harden well* with mercury. When added to an amalgam of silver and tin, however, it produces a decided effect in lessening not only the shrinkage, but also the tendency to ball up and round itself in the angles; but while it does this, *it greatly retards the setting.*"

The Professor of Dental Metallurgy, of University of Pennsylvania Dental Department, states that he found an alloy of *gold one part*, silver four parts, tin five parts,—as simplified from his "two thousand five hundred milligramme" arrangement,—when mixed with half its weight of mercury "retained its sharpness of edge, *hardened well in a few minutes*, and apparently filled all the requirements of a dental amalgam."

* Now eighteen or twenty—1890.

At first sight the discrepancy of these statements is confusing, but a closer examination renders them even more so. Reducing the "parts" of the two alloys to the usual analytical "hundredths," for the purpose of easy comparison, we find them to be, practically,

$$\left.\begin{array}{ll} \text{Gold} & 17 \\ \text{Tin} & 33 \\ \text{Silver} & \underline{50} \\ & 100 \end{array}\right\} \text{ which "}\textit{does not become very hard,}\text{" and in which gold "}\textit{greatly retards the setting,}\text{"}$$

and

$$\left.\begin{array}{ll} \text{Gold} & 10 \\ \text{Silver} & 40 \\ \text{Tin} & \underline{50} \\ & 100 \end{array}\right\} \text{ which "}\textit{hardens well in a few minutes,}\text{"}$$

From an immense experience in the making, testing, and working of amalgam alloys, I should infer that the statement of Prof. Hitchcock was largely based — as he intimates — upon hearsay; while the other statement is evidently based upon actual experiment; and yet the two statements are eminently calculated to mislead, and especially when *contrasted* as we have them above; for it would naturally be inferred that the second formula, in regard to which it is said that, when properly mixed, it "apparently filled all the requirements of a dental amalgam," must be decidedly the better of the two; while the facts are that, when properly mixed, the supposed "English amalgam" is much superior to the other. It "sets" just as quickly; gives as good a "color-test,"— maintenance of color in dilute sulphuretted hydrogen water;— hardens with greater density; permits of finer cutting, and "works" with greater smoothness; has less "shrinkage" and better "edge-strength," and yet *it does not* "fill all the requirements of a dental amalgam."

NOTE.—I *think* Prof. Hitchcock must have been entirely misinformed regarding the alloy to which he has referred; for I do not know of any such material, and I do not think it likely that anything so peculiar — if in the market — could have escaped my notice. The "first cost" of such an alloy would be *at least* four dollars per ounce in the ingot. The addition of the manufacturers' and dealers' profits almost always makes the retail price treble, and sometimes more than quadruple, the amount of "first cost;" and such an alloy would not, therefore, be sold for less than ten dollars per ounce. So far as I know, there is none such advertised.

From our line of experiments it has been concluded that gold, *more than any other metal, in proportion to the small amount required,* diminishes "shrinkage," increases rapidity of "setting," imparts *fine grained* plasticity, aids in "maintenance of color," and secures desirable edge-strength to amalgams; and furthermore, that, like any "balancing power," its great usefulness is developed in exact ratio with the accurate meeting of requirements on the part of the other component metals.

It is for this reason that I have given it as an "undetermined element."

It seems, at present (1890), that the *best* proportionate quantity for gold in alloys for ordinary work is from three to four per cent., while in alloys for "Front Tooth" work it seems with me to have poised, after much oscillation, at about ten per cent. —but the *value* of any definite proportion of gold seems, with every experiment, increasingly to depend upon the accurate proportionment to it of *every other* component.

ANTIMONY. *Stibium.* Symbol, SB. Fuses at 840° F.

This metal has been, and is even yet, used in a few of the many alloys at present offered to the profession with the stereotyped assertions as to *their* "superior excellence." The *only difference* between them and the general alloy consists in the addition of a portion — sometimes quite large — of antimony. It is conceded that the one property of "shrinkage" is controlled by this modification, and that the grain of plasticity is finer; but our experiments resulted only in duplicating purchased alloys which, when made into amalgams, proved so *excessively dirty* in mixing and in working, that, for this reason alone, all further efforts with this metal were abandoned.

Shrinkage is so satisfactorily under control, and fine-grained plasticity is so easily obtained, that, as the working of the best amalgams is yet accompanied with undesirable degree of "soil" to the hands, any ingredient which only accomplished the two requirements above mentioned, and did it with excessive increase in so objectionable a direction, seemed to us unworthy of much consideration.

ZINC. *Zincum.* Symbol, Zn. Fuses at 773° F.

As the alloys of copper and tin have been spoken of as very curious and valuable, so may the alloys of zinc and copper be spoken of as eminently "ancient and honorable;" and when to these must be added the quality of exceeding usefulness, we may well come to the consideration of zinc with befitting care and deference. Not that it has been, as yet, *proven* to be of such lasting value in dental amalgam alloy as to reflect credit upon its "made record;" but that, like gold, its *possibilities may be* something of vast importance.

Added to the usual "40 silver, 60 tin" alloys, in the proportion of from 1 to 1½ parts in 100, it seems to control "shrinkage" perfectly. So decided is this, that fillings made of such amalgam, in tubes of five or six times the diameter of those usually employed in the "leakage test"—with blue or purple ink—give no perceptible indications of permeation of fluid.

Added to such alloys as Lawrence's, Hardman's, Pierce's, — made from analysis of "Standard,"—"Standard," etc., in the proportion of even less than 1 part in 100, the same result is produced. Besides this, it seems to impart an additionally "buttery" plasticity to the amalgam, which gives it exceptionally fine working quality, and also to add to the already satisfactory *whiteness* of the filling, and to its *maintenance of color.*

Altogether, the effects which seem to be due to limited admixture of zinc are expressed to each other, by experimenters, in using the somewhat vague but comprehensive word, *"peculiar."* It is hard to say just what this means, or in what degree in any direction, but it signifies that the "feel" of the "make" is different; that the "amalgamating" is different; that the "working" is different; that it "sets" differently, and that the *final result* is different; and that it is *satisfactorily so.* This line of work is of such comparatively recent date — a few years — as yet to be entirely within the boundaries of "experimental." It *seems* to promise well, but the *intimate relations,* as well as the *distinct differences,* which exist between zinc and cadmium should be remembered. Zinc for alloys *must* be C. P.

CADMIUM. *Cadmium.* Symbol, Cd. Fuses at 442° F.

It is about forty years since cadmium amalgam alloy was suggested to the profession by Dr. Thos. W. Evans, of Paris.

The *promises* of this alloy were certainly very alluring It was easily amalgamated; the amalgam was readily inserted; it did not discolor; it "set" with surprising celerity; it made a sufficiently resisting filling. What wonder, then, that the gentleman who introduced it was pleased with the material? His mention of it, however, was very soon proven to be not in accordance with the slowly, and sometimes painfully contracted *habits* of deliberation, watchfulness, and long-enduring patience which pertain—as a part of himself—to the old experimenter, but must be viewed as entitled to our respect and gratitude from the generosity of his outpouring of that which seemed to him so very valuable an adjunct to the practice of his professional brethren.

My own experiments with cadmium amalgam were disastrous in the extreme. Presuming upon the high authority of its recommendation, I received it with the cordiality of a young enthusiast, introduced a large number of fillings,—nearly two hundred,—and was delighted with it.

My satisfaction was, however, very short-lived, for only three or four months passed before sundry indications presented, which aroused my suspicions as to the uniform integrity and durability of the material,—these were an occasional, but evident crevicing at edges; a gradual softening and disintegration of some fillings; and the yellowish discoloration sometimes apparent in adjoining tooth-structure.

Having kept my usual list for statistics, under the head of "cadmium work," I immediately sent for several patients to examine as to how things were progressing. Strangely enough, so large a proportion of the fillings looked well and seemed to be doing good service, that my suspicions were quieted until the time arrived for the periodic dental examinations which I early instituted, and endeavored by persuasion and by argument to enforce, among my patients.

Then it was that I became fully impressed with the utter worthlessness of the specious amalgam. Quite a number of fillings were found completely "demoralized," and, what was far worse, quantities of dentine had become thoroughly decalcified and stained to a bright orange-yellow color—sulphide of cadmium.

I at once undertook the most thorough examination possible, and commenced the work of reparation of damages. So far as the removal of fillings in *pulpless teeth*, and their replacement by those of less pretending, but more trustworthy materials was concerned, all was well enough; for, in that work, no other suffered nearly so much as I. But of the teeth containing vital pulps, of which I am truly thankful there were comparatively few, the most became "devitalized." There was but a very occasional recollection of pain, and this had been of so slight a degree, as not to have compelled a visit for relief upon the part of any one; but in such cases, as I gradually removed the yellow portions of tooth-bone until I finally entered the pulp cavities, I found more or less thoroughly devitalized pulps. In some instances, where thick septa of dentine existed between the bottoms of the cavities of decay and the pulp cavities, the pulps were still living; in these I carefully excavated until I had removed all yellow tissue, saturated the remaining dentine with creosote, as was then the practice, and refilled.

Some, even of these pulps, have since died under the later fillings; the putrescent pulps have given rise to peridental irritation, and the fillings have either been removed or the teeth have been tapped and treated. Others, again, have lived and are yet living; but *even these*, as I occasionally see them, act as reminders of my woful experience with cadmium amalgam.

But cadmium is yet a component of a few of the present amalgam alloys. It is introduced in some alloys, in very small proportions, from one to three per cent., by those whose experience seems to dictate to them the propriety and the advantage of so doing. It is an open question as to the advisability of even this slight admixture, but for such alloys as have this metal incorporated in large proportions, and even such are occasionally advertised upon the pages of our dental publications, I have but one opinion, and that is, that they should be *denounced with the utmost severity*.

As I have said, the question of the absolute expunging of cadmium from the list of metals permissible in amalgam alloys, or of its retention as one which, in very limited quantity, may be possessed of real value, is yet an open one. *So far as I know*, there are no positive grounds for its retention; there are no reasonable deductions which point to it as of any value. I do not know of any essentials to a good alloy, which are positively given by cadmium, unless in detrimental quantities, that are not attainable by means which are not questionable; and as I cannot but fully appreciate the probable impossibility, on my part, of any further *extended* experimentation, such as I have prosecuted and enjoyed during the past thirty-five years, I desire to caution the workers of the present day as to the need for care in their endeavors for the "positive placing" of cadmium.

While I could not do other than concur with the advisability of *careful, thoughtful, observant,* and, above all, *patient* cadmium experiments, I nevertheless feel that I should be derelict if I did not give it as an opinion, that, though cadmium is a most enticing metal, it is also a *most dangerous* one.

Those forewarned, surely should be forearmed.

PLATINUM. *Platinum.* Symbol, PT. Fuses, alone, only before the oxy-hydrogen blowpipe, or in a very powerful blast furnace.

Although this metal has been so generally thought to be an important one in amalgam alloys, and although its value has been so positively stated and so tacitly acknowledged as to have made its *name* alone, without any known quantities, *or even actual presence*, of sufficient power to bestow position of eminence upon material to which it is given, I have nevertheless placed it *last* upon the list from the fact that, *for years*, it has passed from the notice of our "New Departure Corps," and has been, by us, just as thoroughly ignored as it has been, in public, ostentatiously paraded.

In the proceedings of the "New York Odontological," Dec., 1874, page 52, in Dr. E. A. Bogues' paper on "The physical properties, etc., of dental amalgams," the following is given:

"Fletcher's Platinum and Gold alloy, marked VIII., and yielding upon assay these two different results:

Gold	. .	3.60	Gold	. .	5.10
Platinum	.	3.30	Silver	. .	39.50
Silver	. .	37.63	and Tin	. .	55.40
Tin	. .	55.47	Platinum, *none*."		

The information that a "*platinum* and gold alloy" had as

one of its components "platinum *none*," is certainly amusing; but that "dentistry" should have accepted and *used* such material *because* it was "*platinum* and gold alloy," is assuredly discreditable in the extreme.

And yet in this instance ignorance *was* bliss, and perhaps it is true that "when ignorance *is* bliss, 't is folly to be wise!" for, of the two alloys analyzed, that with the "platinum *none*" is the better alloy! Its 39.50 of silver and its 5.10 of gold gives it quicker setting, better edge-strength, better color, and less shrinkage than pertains to its rival with only 37.63 silver and 3.60 gold; while in each the more than 55 of tin does all it can in the way of decreasing "setting," permitting "bulging," and consequent "crevicing" and diminishing "edge-strength;" surely these things *are* difficult to regulate, even with "judgment."

The conflicting opinions regarding platinum are quite equal in antagonism to those which have been quoted concerning gold.

In "New York Odontological," Dec., 1874, page 20, Dr. Cutler says: "Filings of platinum and mercury rubbed together . . do not amalgamate readily. . . . There appears to be no affinity between the two." "I combined platinum with silver and tin in small proportion, and found that just in proportion to the amount of platinum was the amalgamation retarded." "The next experiment was with platinum and gold, one equivalent of platinum and two of gold; . . . after the lapse of two weeks it was *not firm*, . . . in fact, it *did not harden* to any extent." The next experiment was with one part platinum, two parts gold, and three parts ordinary alloy. "The mass was grayish-white, rather dirty in appearance, and *did not become hard* and *firm*." The next experiment was with one part of *platinum*, one part of *gold*, two parts of *silver*, and two parts of *tin*. After twenty-four hours the lump was "white and pure, but *not hard enough* for durable work;" "in twenty-four hours longer—in forty-eight hours—the mass had become quite *as firm as ordinary* amalgam." This was a mixture of platinum, gold, silver, and tin—the combination which has been stated as *alone* utilizing platinum.

These experiments were not alone made with "alloys," but

were a "line of work" in what we call "mechanical mixes"—done evidently with the idea that such mixing was equivalent to alloying. This is by no means the case, but the work is interesting as showing the effect of platinum when thus "mixed."

Mr. Fletcher says—so states Prof. Hitchcock—that "the amalgam called *platinum-amalgam* is composed of the ordinary silver and tin alloy with ten per cent. fine gold, to which sufficient platinum is added *to cause it to set quickly*."

The quantity required to produce this effect is not given; but it may be well to recall the fact that *without the platinum* the formula is *practically the same* as quoted from the University of Pennsylvania Professor of Metallurgy, which, *minus platinum*, he said, "hardened well in a few minutes."

Prof. Hitchcock evidently accepted the theory of the *quick-setting* control of platinum, for he says of palladium, "When added to a gold, silver, and tin amalgam, it hastens the setting about the same as platinum does."

When we read from Dr. Cutler that the "mechanical mix" of "filings and mercury do not amalgamate readily;" "do not form a metallic mass at all, but remain in the form of a dark powder;" and that "there appears to be no affinity between the two," and when we read from the University Metallurgist that "a very smooth and plastic amalgam may, however, be formed by rubbing some finely divided platinum, such as is obtained by precipitation, with mercury, in a heated mortar;" but that "an amalgam composed of platinum and mercury does not harden well," and that in an alloy of tin, silver, and platinum the properties were "greatly impaired by the addition of platinum;" and that in an alloy of tin and platinum "the property of setting was almost entirely lost," and when all experimenters agree that palladium amalgam "sets" with such rapidity that unless it is made very soft it cannot be properly inserted as a filling, it seems doubtful as to *its* hastening setting "about the same as platinum."

But, again, while it has been usually accepted that, *in some way*, platinum is a very advantageous adjunct in an amalgam, and while to it has been ascribed the power of preventing discoloration alike of fillings and of teeth; the increasing of plasticity; the promotion of setting, and the production of a *generally excellent* (!) filling material, it has been taught by a few that "the real value of platinum" pertains *only* to its combination with *tin, silver, and gold.*

Its "real value" in this connection is stated to be the giving to such alloys the properties of "almost instantly setting" and of "greater hardness."

Even granting this to be true, the "real value" of this metal seems to be of very little, if indeed it is of any, moment; for if, as we know, alloys can be made without it, composed only of silver, tin, and gold which will "set" with even more than desirable rapidity, and if, as I have stated, the softest of all the amalgams are needlessly hard, and the harder ones would be very much improved if they could, without loss of other valuable quality, be made softer, then the "real value" of platinum becomes so questionable as to render it more than probable that it may be *positively detrimental.*

But the line of work done by our "corps" does not corroborate either assertion concerning *much* control of discoloration; *much* control of "plasticity;" *much* control of "setting," or *much* induction of "hardness;" and for amalgam, so far as "general excellence"—vague and unscientific as is the term—is concerned, I believe that *one part* of zinc is worth a dozen times more than *a dozen parts*, or any other number of parts, of platinum.

So far as is *proven*, the value of platinum seems to be *just equal with that of tin.* Every alloy which we made or experimented with, that had, *in place of* certain proportions of platinum, *the same equivalents of tin*, set the same; shrank the same; gave the same color-test; had the same edge-strength; and, with the exception of being, perhaps, a *little less* plastic and a *little less* tough, worked the same; and, so far as could be positively demonstrated, were *practically the same.*

My conclusion is, therefore, that *except in name* (!) the metal platinum is *valueless* as a component of amalgam alloys for filling teeth.

With the added experience of the past five years I have no reason for modifying this conclusion.

ARTICLE VI.

THE MAKING OF AMALGAM ALLOYS.

THE *making of alloys* for amalgam is a branch of dental manufacture the importance of which is but little realized. I took occasion to direct attention to the large proportions which this had assumed, when I read my paper upon "Plastic Filling as a Power in Dentistry" at the American Dental Association, August, 1878.

NOTE.—This paper may be found in the "Dental Cosmos" for September, 1878, page 474. It was *not published* in the "Transactions" of the Association. The reason for non-publication is given in the following letter from the chairman of the Publication Committee.

174 STATE STREET, CHICAGO, April 5, 1879.

DR. J. FOSTER FLAGG, Philadelphia.

Dear Sir.— In explanation of the fact that the paper read by you at the last meeting of the American Dental Association does not appear in the "Transactions," I am instructed by the Publication Committee to say that it was after careful consideration that they unanimously decided not to publish it.

In the Constitution are the following instructions to the Publication Committee: "They shall superintend the publication and distribution of such portion of the Transactions as the Association may direct, or the Committee judge to be of sufficient value." The Committee felt compelled, under the latter clause of these instructions, to exclude this paper, it being mainly a statement of the claim of the writer, of his greater success in saving teeth — by means of a system of practice NOT described — over that of other practitioners.

The decision of the Committee was not actuated by any personal prejudice, nor influenced by any opposition to the doctrine advanced in this paper, but was governed solely by a sense of duty.

Believe me to be, in behalf of the Publication Committee,

Very truly yours.

I there showed that, *although* amalgam was rather decried as a filling material; that it was only used in exceptional cases; that its employment was usually spoken of as derogatory to "first-class" ability, and that the almost universal testimony of speakers upon it was, that they filled with it but "very infrequently," it nevertheless required the united manufacturing capability of *more than twenty* makers of alloys to supply the constantly increasing demand.

Nearly every manufacturer of "dental supplies" has, through "a long and elaborate series of experiments," arrived at the same formula, and each is "superior" to all the others! By most dealers the "alloy" is advertised as an "amalgam," and it is thus designated by those who purchase and use it; the distinctive difference between the two being, by custom, ignored.

Occasionally, some speaker has given his formula and "method of making," and in the aggregate these are singularly harmonious; from first to last *the silver is melted first*, and the tin is then added in pieces, or each is molten separately, and the two melts are poured together.

As platinum and gold became adopted as ingredients, the process continued essentially the same. The usual direction for making alloy is that given by the Professor of Operative Dentistry, University of Pennsylvania,—"Pennsylvania Odontological," March, 1879,— in which, after stating that "much depends upon the proper alloying of the metals," it is still directed that "the silver should be melted first, and when at a boiling heat the platinum should be added in very small particles, either rolled into thin ribbons or cut into minute pieces. Next, the gold should be added; and, *lastly, the tin,*"— the italics are mine.

Even with this periodic dissemination of instruction, the making of alloy seems to have been undertaken but by few practitioners; no education in regard to it has been given in most of the colleges, and not one graduate in an hundred has any definite idea, either of the components, the proportions, or the properties, of the materials he purchases for the filling of teeth, other than as given him upon the printed envelope or the paper of "directions." And all this, notwithstanding the time-honored "annual announcements" from "chairs of Dental Metallurgy."

Such "methods" are not held in high esteem by the "New Departure" metal-workers, for *their method* is so very different.

For the making of alloy, the hessian, or sand crucible, is used. In this is first fused a very small portion of borax, sufficient to fill the roughness upon the sides of the crucible, and thus prevent loss by adhesion of the molten metals. Any ordinary coke or coal-fire is all that is required for the "melt;" but it is, of course, more systematically, and perhaps more readily, done at the usual dental or smelting forge-fire.

Having perfectly fused the borax, and having thoroughly coated the sides of the red-hot crucible by smoothing the fused borax upon them with a red-hot iron rod of suitable size and length, the comingled metals are poured from a narrow sheet-iron scoop into the crucible, and a slight sprinkling of pulverized borax is thrown upon them.

The fusion of metals is now allowed to take place, and it is surprising, even to practical metallurgists, with what celerity the mixture is melted. *The tin melts first,* and hastens the melting of the more obdurate metals in a remarkable manner.

After the metals are *thoroughly* fused they should be as thoroughly stirred with the red-hot iron rod.

When perfectly melted and mixed, the fused mass should be quickly poured into a broad, open, flat, shallow matrix made of iron or soap-stone; this favors prompt cooling, and thus secures the greatest uniformity of distribution to the components.*

I cannot understand how any reasonable method of working metals, even silver and tin, could result in such wholesale separation of the components of an alloy as that described by the Professor of Metallurgy at the April, 1880, meeting of the New York Odontological. I can believe that there might be approximation to the almost complete separation of the silver from the tin, thus leaving one end of the ingot nearly all silver, and the other nearly all tin, by very slow cooling and very careful work *for the definite accomplishment of that purpose;* but I cannot think that any one, in the least degree proficient in "making melts," need ever fear such untoward experience.

On the contrary, I know that novices in the working of metals, "first course" men, are in the habit of making good samples of three and four metal alloys with but comparatively little practice.

The "cutting" of the alloy into "grains," "filings," or "a kind of powder," as it is variously given by the manufacturers, is a matter of grave consideration to them, as it is a troublesome and expensive process, just in proportion to the goodness of the alloy. Thus, a heavily tin, two metal alloy, can be cast into cylindrical ingots, and "rasped" up into "grains," or better yet "turned" up into "shavings," with rapidity and cheapness;

---

* As suggested by Mr. E. G. Quattlebaum, an excellent matrix is made by using Teague's Impression Compound, or a mixture of equal parts of kaolin and plaster of paris.

while with a fine, heavily silvered, four metal alloy, the work is very different, and is both laborious and much more expensive; but for *individual need* it is only necessary that the file be used.

Here again the quality of an alloy is quite accurately tested; for while the coarse files known as "vulcanite files" are best for cutting all "low grade" or "ordinary" alloys, they are not nearly as good as finer files for cutting "high grade" alloys.

It is one of the "tests" for a good alloy, that it shall not, in cutting, "clog" a fine file.

After the alloy is "filed up," the filings should be passed through a fine wire sieve; this removes all coarse pieces, bits of leather from the wire file-brush, and other undesirable impurities which may become mixed with them during the filing. After this the filings should have a magnet passed through them till no steel filings adhere to it. Then, by shaking the filings from side to side in a *flat* tin scoop nine inches long, five wide at the mouth, eight wide at the back and one inch deep, the dust, specks, motes, etc., come to the surface and can be *blown* off, leaving the filings perfectly *clean*.

The filings are now *prepared* for use, but they are, most decidedly, *not ready* for use.

Although manufacturers have not hesitated to cut up large quantities of alloy, and keep it "in stock," without fear of deterioration, it has nevertheless been thought by them, and by the profession generally, that such keeping was detrimental, and from this belief some practitioners have filed up or turned up but small portions of their ingots at a time, in order that their material might always be "fresh," as it has been termed. Among all the many errors which have obtained in connection with alloys and with amalgam, there is probably no other more decided than this. There is no alloy made that does not work better and make better results after it has been cut for several weeks than can possibly be the case where it is "fresh," and in this particular, as in every other, the distinctions are so marked between "ordinary" and "high-grade" alloys, that an "expert" can decide in a few minutes, simply by mixing, not only the *quality of any given sample of alloy*, but, approximately, *how long it has been cut*.

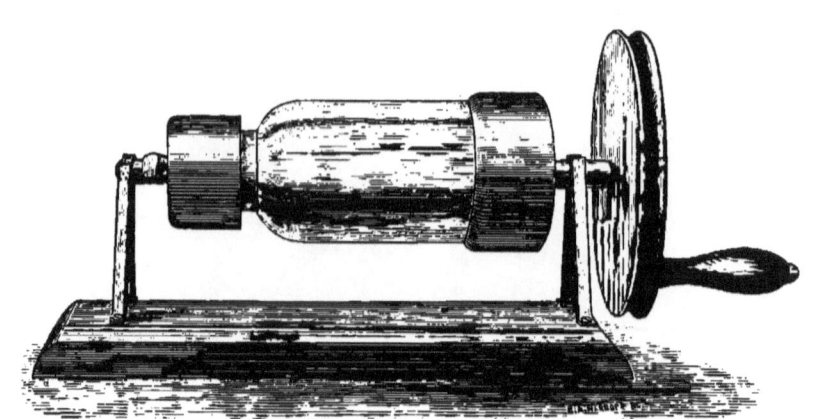

THE MAKING OF AMALGAM ALLOYS.

An alloy *must be poor* that will mix with a small, relative proportion of mercury and work satisfactorily when freshly cut. And inversely I can state that no really good alloy, such an one as will, when properly "aged," rank well as amalgam under the "setting," "shrinkage," "edge-strength," and "color" tests, is fit to work unless it has been cut at least *two months*. In my own practice, I never use any alloy that has not been cut for three or four months, and I prefer them even *very much older* than that; alloys that are "dull of response," that is, which apparently work well amalgamated when freshly cut, shrink notably, set slowly, bulge markedly, and have little or no edge-strength.

During a line of work upon the mixing of various fresh-cut alloys, it was found that, by placing them in a revolving glass cylinder, and maintaining revolution for several hours, two simultaneous effects were produced. Not only were the alloys thoroughly mixed, but a result analogous to that which is given by "time," became apparent. The working quality was much improved, and for this reason the double name of "mixer" and "ager" was given to the instrument which is here illustrated.*

Qualitative Testing of Amalgam Alloys.—The following brief directions, for which I am indebted to Messrs. Eckfeldt and Du Bois, may be found useful and interesting to those who desire a general idea of the method of making some simple qualitative tests of amalgam alloys.

Dissolve from ten to twenty grains of the alloy in a small quantity of nitric acid, say half an ounce, by the application of gentle heat.

The appearance of the solution gives the first clue to its composition; the tin, being undissolved, shows itself in the form of a white precipitate.

Should gold be present, the tin will be colored from a light to a deep purple according to the proportions; suffice it to say that less than one-half of one *per cent.* will make quite a decided purple.

The presence of platinum, with or without the gold, gives the tin a dirty, blackish color, and the platinum being partly dissolved by the nitric acid, the whole solution is, likewise, discolored.

* See Appendix, Sec. 1.

Copper, in quantity, colors the solution green or blue.

These are the first general appearances which determine the presence of tin, gold, platinum, and, possibly, copper.

For the determination of the remaining metals found in these alloys, the solution should first be evaporated to dryness; then dilute with distilled water and eliminate the oxide of tin by filtering.

The silver will be precipitated as a chloride, flocculent, but densely white, by treating filtrate with very dilute hydrochloric acid. Eliminate by collecting on filter.

Copper will show itself by a blue color, if a small portion of the filtrate be now tested by pouring into it a little ammonia. Should this not appear, treat the filtrate with sulphuretted hydrogen. If cadmium be present, it will be shown by a bright yellow precipitate. If the color be not bright yellow, but brownish, it indicates that some silver has either been left unprecipitated by the chlorine or has passed through the filter.

Should the ammonia test decide the presence of copper, treat filtrate with sulphuretted hydrogen as before directed, and a black precipitate of sulphide of copper will result, unless modified in color by a large percentage of cadmium. Collect this copper, or mixed precipitate, on a filter (*preserving the filtrate*); if copper and cadmium both be present, the *precipitate* must be boiled in dilute sulphuric acid; this dissolves the cadmium and leaves the copper to be collected on a filter.

Treat this *last* filtrate with sulphuretted hydrogen, adding a few drops of ammonia to the solution, to modify hyper-acidity. The bright yellow precipitate will be the sulphide of cadmium.

Lastly, boil down the *first* filtrate from copper, or mixed precipitate referred to above, until nearly dry, to expel the sulphuretted hydrogen; then add a little pure water and a solution of carbonate of soda until neutralized. The presence of zinc is then proved by the white carbonate of zinc now precipitated.

ARTICLE VII.

TESTS FOR AMALGAM.

THE various "tests" for amalgam are (1) the "quality" test; (2) the "shrinkage" test; (3) the "setting" test; (4) the "color" test; (5) the "edge-strength" test; (6) the "tooth-conserving" test.

First. *Quality Test.*—The quality of an amalgam depends upon the quality of the alloy with which it is made, the quantity of mercury with which it is mixed, and the method by which the filings are incorporated with the mercury. The *quality* of the mercury has no appreciable influence upon amalgam; it is only needed that it be *pure*, that is, practically free from metallic admixtures; thus, the mercury as sold in the ten-pound stone bottles is perfectly adapted for making dental amalgam, and the necessity for having it "double distilled"—a quality thought to be something finer than is usually sold—is merely ideal. *I* "double distill" by pouring from my ten-pound bottle of "battery mercury" into my box-wood mercury holder.*

Mercury which has been used for a *palm* button of amalgam, and has been "squeezed out" *by the finger*, should not be returned to the mercury holder, as it contains an indefinite amount of various metals in unknown quantities. The presence of these metals, held loosely for a length of time in the mercury, not only prevents the making of a "known" amalgam by its use, but likewise so affects the mercury as to render it a less powerful melter of the other metals, thereby requiring a larger relative proportion of mercury than is desirable in the making of any given amalgam.†

Testing quality is, however, governed by different considerations from making amalgam, though the need for pure commercial-mercury is equal in both directions. And thus it is, that while each kind of alloy requires a definite relative amount of mercury and especial methods of mixing for the attainment of the best filling possible to be made from it, *all alloys* require that *the same* relative proportion of mercury

---

* See Appendix, Sec. 2. † See Appendix, Sec. 4.

shall be used in the making of "test-buttons" that shall decide "quality."

The reason for this is, that it is essential that *one definite proportion* of metal *be known;* and, as the mercury is the only metal to be added, of this must be made the definite addition.

For this purpose, then, *equal parts* of mercury and alloy should be taken,— these proportions have been chosen for the reasons that they are most easily weighed,— the mercury being placed in one scale and the filings in the other, and these accurately balanced; and that no alloy in present general use requires such an amount of mercury for the making of an amalgam fit for filling purposes.

Filling purposes and testing purposes are different; in that for filling, rapidity of accomplishment of purpose — within the bounds of obtainance of excellent result — is desirable; while for testing, reasonably slow progress better demonstrates relativity of consecutive showings.

Having weighed the portions of mercury and alloy to be tested, the method of mixing is practically immaterial. The mix may be made in the palm of the hand; it may be a "shaken-mix,"— as suggested by Mr. Fletcher,— pressed into form by his mould; or it may be the usual "mortar-mix," with the subsequent palm-kneading. In no case, however, should any of the mercury be removed from the mass. Thus, if the mass be pressed in a Fletcher mould, it should be allowed to retake the mercury squeezed out in the packing.

By any of these various methods of mixing, "buttons" of different degrees of plasticity are made which are in known accord with the varying composition of alloys.

Buttons made from alloys of tin and silver — largely tin — are very soft, having the *peculiar* plasticity of *tin;* these "set" slowly; gain edge-strength with the utmost deliberation; *never* attain an acceptable edge-strength; retain for a long time — an hour *or more*, according to the relative quantity of tin — a degree of exterior softness which permits of marking by rubbing gently with the finger, and are easily crushed or broken after several hours of hardening.

Those made from alloys of silver and tin — largely silver — are much firmer in consistency; have the peculiar crepitation, indicative of silver, in making; "set" much less deliberately, gain edge-strength more satisfactorily; attain fair edge-strength in from two to four hours; and lose in an hour that exterior softness which permits of "finger-marking."

Those from alloys of tin, silver, and copper — largely tin — have the peculiar "feel" given to alloy by copper. This may well be called *peculiar*, and yet, to one who is a decided "expert" in this method of analyzing, it is very cognizable.

NOTE. — Some years since, Dr. Weston — of Weston's alloy— was calling upon me in relation to these, to us, very interesting matters, when a sample of alloy, which I had requested a friend to obtain, was handed to me. I opened the package, and giving Dr. Weston some of the alloy, requested an "approximate analysis." Rubbing it up with the palm-mix and gently manipulating it for a few minutes, he remarked, "It's a little *peculiar*. I think it is pretty nearly equal parts of tin and silver, with *-er* a little *-er* copper in it — say about four or five parts in a hundred." It was Lawrence's alloy.

Test amalgams of these alloys "set" less slowly than those of tin and silver of this grade; but the "setting" due to copper is "grainy," not firm, and is accompanied with a characteristic *whitening* of the mass. They remain softish for quite a time, — nearly an hour, — and their edge-strength is not good. According to our experiments, these *begin* to represent "good alloys."

Alloys of silver, tin, and copper — largely silver — are increasedly satisfactory in all directions: they set with increased promptness; they harden satisfactorily and with firmness; they have reasonable whiteness; they have desirable edge-strength; in an hour they have all these test requisites in *test sufficiency*.

Alloys of tin, silver, and gold — largely tin — are those which first give *apparently* "excellent" testings: they work with a pleasant plasticity; they harden reasonably well and with reasonable firmness; their edge-strength seems to reach "quite satisfactory" in quality; if they are cut with fineness, in the form of thin, light, shavings, bulky in mass but diminishing most notably on amalgamating, the mass sets more promptly, and all these qualities are particularly noticeable to those who, as a rule, have been *habitually* working *the tin and silver — largely tin —* alloys.

It is due to these facts that testimonials, honest in their innocence, are found attached in great numbers to the advertisements of this *tin*, silver, and gold class of alloys. The givers have used alloys "comparatively infrequently;" they have used them upon the recommendation of their makers, and with no knowledge of components or proportions. They have learned to recognize *certain peculiarities* of make, set, hardness, and capability of finish, as characteristic of "amalgam," and the marked differences, which *even* a little gold is able to confer upon *even* the stereotyped alloys, appear to *them* as "*something extraordinary.*"

When such alloys are subjected to "general testing," their deficiencies become very decided, and they are proven, in reality, to be but little better than their tin and silver predecessors of the amalgam "middle age."

Buttons made from alloys of silver, tin, and gold — largely silver, but sufficiently of gold — *are* the beginning of *something fine*. They mix with combined rigidity and plasticity; their crepitation is that due to the combination of silver and gold, "short, sharp, and decisive," and peculiarly pleasing to the initiated; their setting is prompt, firm, and dense; they evince excellent edge-strength; and at a time when the buttons of amalgam from *tin* and silver can readily be scraped away with the finger-nail, the buttons of *silver*, tin, and *gold* can be as readily *handsomely burnished.*

I have had the pleasure, frequently, of demonstrating these things, and from expressions of developing convictions, great satisfaction, and marked degree of interest, I have been led to conclude that, if this line of work is done by one familiar with it, no other is capable of so thoroughly impressing the student with the wonderful difference between alloys, the absolute need for "knowledge" in this connection, and the perfect facility with which, *by its possession*, one can separate the "tares" from the "wheat."

In the quality testing of alloys of silver, tin, gold, and copper, or of those of tin, silver, gold, and zinc, the work is much increased in difficulty and delicacy. To do this with any approximate to *accuracy*, it is necessary that one shall have made hundreds of makes of *known* different alloys; and even then it can only be "inferential." But with this experience, a general idea can be formed upon which can be predicated, more surely, the fine work of analyzing such materials, and the finer work, still, of suggesting such modifications as would *probably* be improving.

Here is a field, indeed, for the thought, energy, and discussions of the "coming dentist,"— a field which, though it has yielded fine results, has been but comparatively superficially cultivated. While I think that the workers of "The New Departure Corps" have done dentistry solid service in the labors which they have accomplished, I yet feel that each member of that band now, more fully than ever before, recognizes that, in the aggregate, the task performed is, to the task yet to be done, as but a little part.

In testing the quality of such alloys as contain cadmium, the presence of this metal is usually indicated very promptly upon amalgamation; in proportion of *three* or *four* per cent. the "cadmium feel"— a sort of slippery stick or greasy catch — is just sufficiently apparent to excite suspicion; it is also dirty in its working, and leaves much soil closely adherent to the fingers; it also causes "cadmium setting" in sufficient degree to be noticeable. This setting is rapid but devoid of strength. Antimony gives some of these peculiarities, but cadmium is detected with reasonable facility after it has been *experienced* a few times in testings.

Zinc, as a component of amalgam, is not brought into special notice by the "quality test," except as it seems to *cause adhesion of the mass to the pestle* during making; this is so marked in alloys containing zinc in from two to four *per cent.* that it is well worthy of note.

Second. *Shrinkage Test.*— I believe that to Mr. John Tomes first occurred the idea of testing the bulk integrity of amalgam; his work upon this point was done about thirty years ago. So firmly had the belief in "expansion" of amalgam become impressed upon the dental mind, that, notwithstanding the conclusive proof given by Mr. Tomes in favor of "contraction" or "shrinkage," as it is variously designated, as shown by the whole line of *tin* and silver amalgams, *dentistry* continued to believe, as it had been taught before, that amalgam *expanded in its setting.*

Mr. Fletcher, of Warrington, England, followed in this line of experiment with his practical "tube tests." Of these I have done an immense number. The advantage of the "tube test" over that of the "amalgam micrometer" is the possibility of doing a number of experiments in a limited time, as tube after tube can be packed and set aside for results, while with the micrometer each packing has to remain in the instrument until the result is obtained. This requires from several days to many months for each experiment; but while the tube test is practical, *reliable* micrometric work is more accurate, and particularly as regards *relativity* of contraction or expansion, and is therefore *essential* to *progressive* experimentation.

NOTE.— I must agree with the experience given by Dr. Geo. B. Snow, "Odontological Proceedings," Dec., 1874, page 67, in that I have never been able to obtain the peculiar *mirror-like* appearance which I have seen in some specimen tubes sent to this country by Mr. Fletcher, even though I have repeatedly tube-packed amalgam made from his alloys mixed both dry — powdery — and with sufficient mercury for plastic working.

For the introduction of the "index" amalgam micrometer, I believe we are indebted to Prof. Hitchcock; but the work done by him with his instrument, as presented at the New York meeting, while so very questionable, upon the slightest examination, as to make it seem incredible that it should have been presented in good faith by any one to any scientific body, is, at the same time, *in a certain sense, instructive*, and may, by giving it only a passing notice, prove of value as a warning to other dental organizations, before which such papers might be read in future.

The whole series of experiments are so loosely done; so completely inharmonious in their relation the one to the other; so *impossible of production*, if done with an instrument of the least pretensions to accuracy, in the hands of an experimenter of even limited experience, that no value whatever, as bases for amalgam alloy work, can attach to them.

I have already stated that the analyses are unreliable so far as regards practicality, and have shown that the *marketed alloys* of the same names are not composed of the metals or proportions there given — with one exception; but I shall have to now *suppose* that the analyses were correctly made from samples furnished by the various manufacturers — *a thing which I am sorry to have to do* — in order to prove the want of accord between the results as "*shrinkage measurements.*"

Alloys made of the metals given, in the proportions given, *do not relatively "shrink," as stated*, whether mixed "dry," "medium," or "full;" whether "washed" or "not washed." Two or three examples will show this:

"Arrington's" Silver . . 40 Expanding component.
Tin . . . 60 Contracting component.
"Shrank .0045."

"*Townsend's Improved.*"
Silver . . . 39.00 Expanding component.
Gold . . . . 5.31 Eminently *preventive* of shrinkage.
Tin . . . . 55.69 Shrinking component.
"Shrank .014." (!)

That is, an alloy with 44.31 parts of expanding and non-shrinking components, and 55.69 parts of shrinking components, shrank *three times as much* as one composed of only 40 parts of expanding components and 60 parts of shrinking components!

Again, as if in more complete derision of this famous "shrinker," — Townsend's Improved — "Walker's," with an analysis which places it *for shrinkage* in exactly the same grade of alloys with "Arrington's,"

Silver . . . . 34.89 ⎫ 39 parts expanding and non-shrinking; 1
Gold . . . . . 4.14 ⎭ neutral,
Platinum . . . .96 ⎫ 60 parts contracting,
Tin . . . . . 60.00 ⎭
"Shrank" *only* .002. (!)

But, strangest of all, "Johnson & Lund," with an analysis of

Silver . . . . 38.27 ⎫ 39 parts expanding and non-contracting;
Gold . . . . . .81 ⎭ 1½ parts neutral,
Platinum . . . 1.34 ⎫ 59¼ parts contracting,
Tin . . . . . 59.58 ⎭
Shrank .001+.

Thus, this alloy, which expends only 16 *cents* — per ounce of alloy — for *gold* to control the shrinkage of *its* 60 per cent. of tin to the *minimum* (.001+), handsomely exceeds "Walker's" control of his 60 per cent. of tin, at an expenditure of 94 *cents* — per ounce — for gold; and leaves "Townsend's Improved," with its determined control of only 55.69 per cent. of tin, by an expenditure of $1.06 — per ounce — for gold, so far in the rear that the difference can only be viewed in contrast.

It seems to me that further comment on such work is needless. And yet the *idea* of the index micrometer is, to a certain extent, a good one.

The "index" instrument which I use for "shrinkage" measurements was made by Mr. Henry Coy,— the well known HC, — whose monogram, upon his excellent make of dental goods, so many years duplicated that of S. S. W., and was presented as a contribution from him to the *armamentorium* of experimental alloy work. It was made from the copy of Prof. Hitchcock's instrument used by Dr. Bogue, and loaned me by him.

By suggestions of Mr. Coy very decided improvements upon the original micrometer were made; the great desideratum of accuracy of packing is much more perfectly accomplished; the shape of the circular end of the short arm of the "pointer" was materially altered, as the circle of the original pointer favored largely the heavy shrinkers and condemned unsparingly the light shrinkers; a sliding matrix-slot was so arranged that the ingot could be easily and safely removed at conclusion of experiment, that other experiments pertaining to density and strength of alloy might be performed; and in this improved condition the instrument is well represented in the annexed illustration.

By this instrument, it is shown that there *is* relativity between composition and shrinkage; the heavily tin alloys make amalgams which shrink most notably. If to these alloys gold or copper is added, the shrinkage is lessened in approximately just relation with the amount of non-shrinkers introduced!

But it is to the measurement of the amalgams made from the heavily silver, tin, and copper; and silver, tin, gold, and copper alloys that we are now most attracted; for "shrinkage" in connection with these is reduced to "team rifle-practice."

For this purpose, I now use the two-inch matrix — micrometric — and direct microscopic measurement.

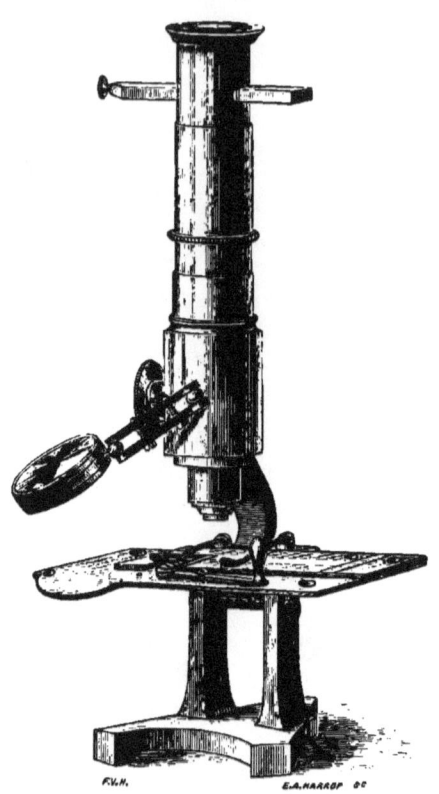

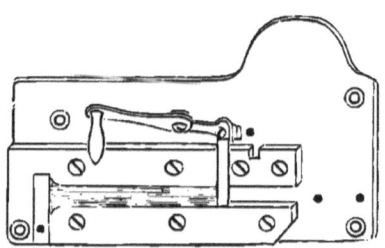

To one of Queen's Household Microscopes the following alterations were suggested by, and made, under the direction of Mr. E. Pennock, to adapt it to this service.

The stage is altered by elevating the spring-clips to take the plate containing the matrix with the amalgam ingot to be tested.

A one-half-inch object-glass of wide aperture, which is better suited for examining opaque objects, is adapted to the instrument.

A Jackson eye-piece micrometer is fitted to the eye-piece, and a stage micrometer, ruled to $\frac{1}{100}$ths and $\frac{1}{1000}$ths of an inch, is furnished for obtaining the value of the divisions of the eye-piece micrometer. The latter is adjustable by a screw, for the purpose of bringing one of the lines of the micrometer to coincide accurately with the margin of the space to be measured. This is, of course, between the free end of the amalgam ingot and the sliding matrix-slot.

By this instrument and arrangement fine work can not only be done in its aggregate, but the gradation of shrinkage, until completion, can be watched and noted with reliable accuracy.

Third. *Setting Test.*— For this test, it is of importance that the exact amount of mercury best adapted to the proper working of each given alloy be ascertained. Each sample is then mixed and worked just as it would be in the filling of a tooth. The object is to obtain the *possibilities* of each amalgam; but in my present amalgam experiments I have abandoned the "dry-mix"— as it is called— and also the use of warm instruments for the introduction of the filling; but any expedient is permissible for the hastening of "setting."

The greatest difference conceivable exists in the behavior of different amalgams under this test; for with some—notably those made from the silver, tin, and gold alloys—teeth can be "built up" in such wise as that they will sustain clasp-work in from thirty minutes to an hour; while with amalgams of mercury, *tin*, and silver, such manipulation can hardly be performed as will permit, *first*, of the immediate "building up," and, *second*, of the utilizing for clasps, in less than several hours, if, indeed, in less than a day, with perfect safety.

In mere experimental testing for "setting," the "set" is regarded as accomplished when the filling will take, *without necessarily retaining*, a bright, smooth burnish.

Fourth. *Color Test.*—By universal consent, I believe, sulphuretted hydrogen water is used for this test. In my experiments I use, for quick results, this water as it is employed in ordinary metallurgic work — full strength; but it is much better, and, in fact, essential, for the determination of *gradation of discoloration* of amalgams, that the tests be made with *dilute* sulphuretted hydrogen water, 1 part to 2, 4, or 6 parts of distilled water.

In such dilutions the *gradual* discoloration of the sample is accurately noted.

Some very strong sulphuretted hydrogen tests, such as solutions of sulphuret of potassium, etc., have been suggested, and have been represented as conclusive provers of "superiority" in amalgams. This opinion is not only erroneous in theory, but is very deceptive in fact. So far from being a test of superiority and goodness, *a powerful sulphuretted hydrogen "color test" is one of the best tests for general inferiority and badness.* Any amalgam which will *not discolor at all* in this test should be viewed with much suspicion, and, indeed, it may be regarded as convicted — upon *prima facie* evidence — of some serious deficiency. The *presumption* is that the amalgam is "cadmiumed;" and that just in proportion as it retains good color, so is it *loaded* with this *tooth-destroying* metal.

It is very important to know that rather *inverse* to the good maintenance of color is the *real value* of amalgam.

In cases where cavities can be nicely "lined" either with good varnish or good oxy-chloride of zinc, it is better that an amalgam of good maintenance of color be used; but *if the preservation of the tooth,* as in a *lone molar* for clasping purposes or for mastication, be the *prime consideration,* all experience indicates that an amalgam *which will discolor,* better accomplishes this end. Therefore, in such cases, and particularly if there are "submarine" complications, alloys of silver, tin, and copper should be used in preference to those containing gold, platinum, zinc, or — worst of all — cadmium.

Fifth. *Edge-strength Test.*—For the suggesting and devising of this important "test," I must claim for myself the original

ity. The necessity of good "edge-strength" is recognized in connection with all filling materials except gutta-percha; but in no other material in which it can be regarded as mainly a *physical characteristic*, is it so important as in amalgam.

In the working of tin, a certain degree of edge-strength can be given by superior manipulation. In the working of gold, a marvellous degree of edge-strength can be given by superior manipulation; while a sufficiency of edge-strength can be given for a majority of fillings by very ordinarily good manipulation. The oxy-chloride of zinc-fillings are not expected to have any sustaining degree of edge-strength, nor to retain for any great length of time even the edge which is first obtained. The zinc-phosphates, though capable of possessing in considerable degree this desirable attribute, are quite deficient as at present in the market. Gutta-percha is accepted as possessing, practically, no edge-strength. Amalgam, therefore, is the *only* "plastic" which can to any extent be depended upon for edge-strength.

It is this fact, superadded to the recognized need for the quality, which made *work* upon this point of such serious import to the "plastic-fillers."

In this, perhaps, more than in any other direction, was it necessary that experimenting should be done with the attainment of results which should be unquestionable.

The usual method of making "buttons," and testing the edge-strength by breaking them with the thumb-nail, was regarded as evidently unscientific and unsatisfactory, notwithstanding its modicum of sturdy practicality. The instrument shown in the illustration was, therefore, devised.

Ingots of amalgam for "edge-strength tests" are made of the definite shape and size given by the bevel-edged matrix. Into this matrix samples of the different amalgams are packed as though for fillings; endeavoring to give to each the greatest edge-strength of which it is capable. After allowing the "test" to *harden thoroughly*, it is taken from the matrix and screwed firmly in position by the thumb-screw lever. The chisel shaped punch-rod is then rested upon the edge of the sample, and the rolling-pea is slowly pushed out upon the graduated scale-beam. The indicator shows the point at which "crush" or "fracture" occurs; and thus is graded the relative "edge-strength," with its accompanying "toughness" or "brittleness."

By this test is also graded the *relative ratio of "setting"* within the bounds of "burnish possibility." Thus, for example, a sample of "Townsend's" is packed and allowed to set for fifteen minutes; it is then carefully removed and "tested" with a light rolling-pea. A sample of "Hardman's" is packed and tested also, after having set for the same length of time; then another sample of "Townsend's" is allowed to set for thirty minutes, and is then tested; and a duplicate of "Hardman's" is also tested after thirty minutes' setting. It will be seen that by this method gradations may be obtained which will definitely "rank" any amalgam, and that, too, beyond reasonable probability of error.

Sixth. *Tooth-conserving Test.*—This, after all, is the grandest test of amalgam. It is the test which can only be done *in the mouth;* it is the test which has been given it in the most trying manner, asking of it to "do" when everything else had *not done;* asking of it to "stand" when everything else had *fallen;* asking of it to "save" when everything else had *failed!*

And how has it responded? If *I* were to say, "nobly!" I might, very truthfully, be regarded as an interested witness, for *amalgam* has largely contributed to my success in the difficult task of *saving everything that had been hopelessly abandoned!*

So I will not testify!

Some of the blackened "shells of teeth" that were filled with coin amalgam more than forty years ago, testify. The thousands of teeth that have been filled, *apologetically*, testify.

The hundreds of teeth that have been filled with the assertion and *the full belief* that they would "last only a year or two," and have lasted *ten* and sometimes *twenty* — testify.

Dentists also testify. In the face of opposition; in spite of vituperation; in contempt of malediction; in defiance of threatened professional ostracism, dentists have testified. Admissions of deadly antagonism, without one particle of knowledge or experiment, have been wrung from some — that is testimony; the possibility of its utility "in certain cases" has been admitted by others — that is testimony; good words have been ventured for it, *slantwise*, ever and anon, even by respectable members; it is believed by them "to be *valuable* as a filling material *when used with proper care and discrimination*"!!! *that is testimony.*

But others have spoken *heartily*, out of their love for truth, out of their *knowledge* gained by years of experiment and *observation.*

Dr. S. P. Cutler says, "Take any mouth or any number of mouths, with decayed teeth in every and all stages of decay, and let any first-class operator fill all decayed teeth on one side of the mouth with gold, and all on the other with best quality of amalgam, say all back of cuspids, the same skill and care being used in both, and await results."

"In my opinion, *based upon observation*, the side of the mouth filled with amalgam, in ten or twenty years, will be found in a better condition than the other."

Dr. C. C. Allen says, "If my life and fortune depended upon the *saving* of a tooth merely, without regard to its appearance, *I would fill it with amalgam.*"

Dr. J. Washington Clowes says, "Thus imperfectly have I discussed amalgam. Conceived in weakness, brought forth and nurtured by empiricism, traduced, maligned, denounced by professionals in high places, malused, abused, and buffeted, it has come to be a power in the land." "Out of weakness it has grown strong, and the very vices of its origin are covered by the mantle of its virtues." "The combination of metals known as amalgam is singular in this: That of all the faults it is said to possess, of all the harm it is said to have done, it is as free and innocent as the child unborn."

"I am impelled to take up its defense at this time by a sense of duty which I owe to it and to every people, wherever in the wide world the voice of its calumniators has been heard, believed, and had power to alarm."

"I know of no worthier act than the performance of duty, no higher aspiration than the enunciation of truth, no attainment more eminent than professional excellence. By the sense of duty well performed, by the truth fearlessly proclaimed, by the excellent idealized and labored for, I beseech you, brethren be not henceforth faithless, but believing."

This is the sound of the advocates of amalgam. It pours forth from the heart with the earnestness of conviction; it comes from the brain stamped and milled as the coinage of intelligence; it has the "ring of the true metal," while the "valuable, if used with *care* and *discrimination*," has the feeble tinkle of the truckler to professional prejudice. Amalgam has *proven itself* to be a notable tooth-saver.

ARTICLE VIII.

PREPARATION OF CAVITIES.

THE very frequent occurrence, in discussions, of the statement that "cavities are prepared for amalgam with the same care *as for gold*," is significant of the fact that *gold preparations* are esteemed as essential to the highest future integrity of the filling. In fact, the remark is usually made in that semi-apologetic tone which seems to intimate the presumption that *for amalgam* the preparations might be thought to be made, ordinarily, after such manner as would be in consonance with the probable inferior manipulative ability of the "plastic-filler;" and it is desired that fellow-members shall be aware that, although the filling to be introduced is *inferior* to those generally inserted, the "preparation" is of that *superior* order usually indulged in by the speaker.

The worker in amalgam recognizes that every principle which governs the preparation of cavities for gold fillings is ignored in the preparation of cavities for amalgam. He, therefore, would not prepare *his* cavities for amalgam with the "same care as for gold" any more than he would prepare cavities for gold with the *same care as for amalgam*. He prepares his cavities for gold fillings, with care, *for gold*, and his cavities for amalgam fillings, with care, *for amalgam, without any thought of comparison of the one with the other*, because there is nothing in common between the two preparations.

In preparing for gold, the governing principles result in the making of free ingress to cavities; of flush walls with antagonizing bearings; of retaining grooves and pits; of cavity shape *triflingly larger at mouth than at base;* and in the absence of undercuts and overhanging edges; all is done to the end that the packing of the gold shall be accomplished without danger of interstice between filling material and cavity wall — *leakage* is the bugbear of the gold-filler.

In preparing for amalgam, the governing principles result in the making of cavities without angles; with no flush walls; few, if any, pits; with cavity shape *decidedly larger inside than out;* with concave undercuts and largely overhanging edges; in short, he aims to make his cavity a *concavity* to the extent of his ability. Why is this? Because he recognizes the spheroiding tendency of his material; he recognizes that it "draws" from angles and from straight walls; he wants to shape it and to place it *as it wants to stay. Bulging* and *crevicing* are the bugbears of the amalgam-filler.

In the preparation of cavities for gold, it is taught to remove all decay which is possible, compatible with safety to the pulp. This is done for the double purpose of a firm foundation upon which to rest the filling, and for the insuring of solid packing to prevent leakage. So important is this to the success of gold work, that *immediately* upon the acceptance of the "leaving of decay" as essential to the *proper* preparation of certain kinds of cavities — a practice which was fought almost as bitterly as was amalgam, but which Prof. Arthur pushed to discussion and acceptance with all the energy of strong conviction, — such devices as golden arches made of pellets, concavo-convex disks of plate, etc., were suggested as *supports* for condensation.

In the preparation of cavities for amalgam all this is lost sight of — the retaining edges, the deep curves, the "holding power" of *properly prepared periphery*, make the mind of the amalgam-worker easy on the score of plug-retention, and he, truly, bestows as much attention, care, and thought upon *not removing decay* as the gold-worker does *upon* its removal!

In a recent society discussion upon amalgam, it was asked, "What is the theory of the action of amalgam upon the teeth?" The reply of one of the most distinguished of those present is reported to have been that "Teeth are rendered calcified by it," and are "saved by the deposits of lime salts from the fibrin in the tubules. I have never seen this excessive hardening under gold." This is unquestionably a statement of facts. Decayed dentine is recalcified, *very usually* under amalgam; and, although I have frequently seen exceedingly hard recalcification — very dark — under gold, yet, the characteristics of the recalcified tissue are sufficiently dissimilar to warrant me in saying that I have *never* seen *such* calcification under gold as I *usually* find under amalgam.

With the S. B. Palmer theory of "compatibility," this seems to us, of the "New Departure," as capable of being reasonably commented upon beyond the confines of mere statement of result. Agreeing with Prof. Arthur, that, "in a majority of cases, *if the cavity be so filled as to preclude* leakage, caries will not progress even though *decayed, dead,* and *decomposed* dentine be left therein"—I regard gold as a means for the arrestation of decay *by this method.*

In teeth of good structure it precludes "leakage;" it acts as a mechanical barrier to the continuance of that decalcification of dentine which has been proven to be beyond the power of the pulp to prevent; this being done, secondary calcification takes place. If the matrix — decalcified dentine — is light in color, the recalcification differs but slightly from original dentine; if the matrix is discolored,—and I have left it sometimes very dark,— the recalcification is subject to that control, and I have seen it almost as dark as under some amalgams.

In regard to relative "hardness," I have nothing positive to say; for it is comparatively infrequent that I have to cut recalcified dentine; but, in my experience, it never has occurred to me to grade recalcification under amalgam as *harder* than that under gold. I have seen it soft under each, and intensely hard under each.

But, as I have said, I have never seen *such* recalcification under gold as I usually find under amalgam. It is this which *demonstrates* one of the *greatest* differences between the two materials as *filling* materials. Take an *ordinary* cavity in a *very soft* tooth: the preparation is easily made; the cavity walls are thick and sufficiently strong; either material is well inserted by any one possessing an average amount of skill.

PREPARATION OF CAVITIES. 83

Both fillings look as though they would do good service, and here the equality ends.

The gold, not liable to tarnish, maintains its appearance of integrity — *it is a thing of beauty;* the amalgam, liable to tarnish, gradually loses its whiteness; it gradually becomes more and more discolored; the surrounding tooth-structure partakes decidedly of this discoloration; the entire tooth is "shaded;" it is *not* a thing of beauty! But what is the result? Prof. Arthur says, "If leakage is prevented, etc.," but in a *very soft* tooth leakage *cannot be prevented;* it comes through the soft tissue of the tooth itself, and moisture finds in gold *an unchanging substance*, resisting all approach, an "incompatible"— according to our ideas — offering *everything but help* in this emergency. It usually is not many years before an examining probe can readily be passed between the filling and the softening cavity walls.

So, in the *very soft* tooth filled with amalgam, the leakage cannot be prevented, but, coming through the tooth-tissue, moisture finds itself in contact with a material *susceptible of change;* a material which, instead of resistance, offers decomposing yielding to approach; less "incompatible" at first, according to our ideas, from the metallic standpoint, increased "compatibility" results from the gradual formation of soluble salts of silver, tin, and copper; these being dissolved, are taken up by the contiguous dentine, which, *with its incorporated metallic salts,* becomes so in affinity with the amalgam filling, with its film of "tarnish"— metallic salts—as to insure almost completely harmonious apposition of tooth-bone and filling, cessation, practically, of decay, and recalcification, *with metallic lustre,* of decalcified dentine.

This, the amalgam-worker counts upon; he requires no firm substratum as a solid resting-place for filling, but places his plastic material, gently, in perfect apposition with the soft dentine, having faith to believe that in due time *both will become hard,* and having some definite *theory* upon which to base the probable realization of his hopes.

In order to insure success, the amalgam-worker largely confines his labor to the peripheral portions of his cavities; in his obtunding of sensitiveness, he mainly bestows attention upon the sub-enamel membrane; in his cavity shapings, his thoughts are always upon spheroiding, and by these means having made preparations in point of fact, from the manipulative, the therapeutic, the theoretic, and, by this combination, from the *scientific* standpoint — equal to anything which could be done for gold — he, in addition to all this, is *helped* by his filling material, and *thus* the *record is made* that teeth *can be saved with amalgam* which *cannot be saved with gold.*

ARTICLE IX.

THE MAKING OF AMALGAM.

IN the earlier days of this material, it was the generally accepted method of making, that the desired portion of coin filings should be placed in the palm of the hand, and should have mercury added to it in sufficient quantity to admit of the forming of a plastic mass by kneading it with a finger of the other hand. To do this properly is a prolonged and difficult piece of work. If an insufficient quantity of mercury was first added, and the mass, in consequence, was not plastic enough for proper kneading and for working as a filling, more mercury was added, in small portions, until the mass was of the desired consistence. If too much mercury was, at any time, found to be in combination with the filings, a part was squeezed out by pressure from the finger. In this manner was amalgam made for its first twenty-five years.

It is now about thirty or thirty-five years since the mode of making which has been referred to, and described, as the "mortar" and "washing" method, was brought to the general notice of the profession by Prof. Townsend. It is not claimed that the ideas originated with him, but, on the contrary, it is admitted that the formula for alloy and the process of "washing" were both given Prof. Townsend by Dr. Hunter, of Cincinnati. I do not know whether these originated with Dr. Hunter; but it has been stated that others were then using the so-called "improved" formula and method.

It was but few years — three or four — before the idea of

alcoholic cleansing was so forcibly attacked that it largely lost its *prestige*. As early as 1859, it was suggested, by Dr. James E. Garretson, that chloride of zinc should be added to the filings and mercury in the mortar, and that these be rubbed up together and the chloride of zinc be then washed out by water. This made a beautifully white filling, which retained its color in many cases for quite a long time, and which seemed even more preventive of tooth discoloration than of tarnish to the face of the filling.

For the reason that the thorough removal of the peculiarly disagreeable taste and "feel" of the chloride of zinc was difficult, I instituted a series of experiments — *reported* in 1861 — which I concluded by the adoption of chloride of sodium (common salt), used practically, as was the chloride of zinc — washed out by water.

As my experiments progressed in the direction of *silver additions* to alloy, the "washing" became less effective in preventing discoloration, but cotemporaneously with this the demand for "lining" cavities increased, so that in the course of ten or twelve years I had abandoned "washing" as *needless*, and had instituted "lining" with oxy-chloride of zinc quite universally, especially in front teeth. It remained, however, for Mr. Fletcher, of England, to demonstrate that "washing" was absolutely *detrimental*, as it produced a condition of amalgam which greatly facilitated and increased leakage.

Immediately upon his announcement of this fact, I made a line of duplicate work upon this matter, and found that my results *corroborated his conclusions most positively*. After washing, either with alcohol or water, the amalgam would permit leakage in an extraordinary manner, both quickly and in large degree, and this, too, although the mass was not only ordinarily dried with a napkin — as was the custom — but even when it was attempted to make a *thorough* drying by using several *hot* napkins consecutively. It therefore is concluded that "washing" — so far from being useful — *is positively detrimental*.

In regard to the *mortar-make*, my conclusions have been very different, for with all changes, and having tried all suggestions modifying this, I yet adhere to it. In Article V. reference has been made to the proposed mixing of alloys with the smallest possible amount of mercury; that thus the amalgam be made "dry," as it is termed; that the result be in the form of powder, and that it be then placed in cavities and rendered plastic by working with heated instruments. As this method is almost impossible of practice in upper teeth, and particularly in difficult and inaccessible cavities, the ingenious devices of, *first*, amalgamating by succussion — placing the filings and small proportion of mercury in a little bottle and shaking them together into powder — and, *second*, moulding the powder into pellets by pressure, using for this purpose a small cylindrical matrix and plunger, were suggested by Mr. Fletcher.

For the requirements indicated, these methods are certainly the best that have ever been offered, but their *necessity* is based upon the fact that the alloys to be so treated are most largely composed of *tin*. With this ingredient, it has been shown that slowness of setting, loss of edge-strength, and tendency to spheroiding, have all to be combated, and this work is, unquestionably, aided most materially by using a minimum of mercury; but, as in the accomplishment of this purpose *by this method*, an amalgam — even in pellet form — of such consistency as to necessitate its manipulation by heated instruments is afforded, and inasmuch as every such complication is disadvantageous from the operative standpoint of mutual comfort, it has been essayed to arrive at equally good, and if possible better, results by methods which will permit of *desirable plasticity* and of *cold working*.

Furthermore, it is contended, and, from experience, I think with reason, that an amalgam is *better* — more thoroughly an amalgam, more homogeneously plastic, tougher, capable of being more readily and more properly manipulated, and capable of producing better "testings" — if amalgamated upon the principle of *mercury melting at ordinary temperature*.

In this process, mercury is added to the filings in sufficient quantity to make a plastic mass which shall meet the varied requirements of "plastic-filling." One of the most important of these requirements is that of *easy introduction;* and whatever else may be accomplished by the "powder" form of amalgam, this certainly is seriously interfered with. But, again, the other requirements of *quick setting*, non-shrinkage, strength of edge, and maintenance of color, are all *essentials* to the æsthetic worker; and it is for the obtaining of these that the *fine work* on alloys has been done and is continually progressing. As I have previously intimated, those operators who have used such alloys as Hardman's, Lawrence's, and Standard have found *no need* for powder-like amalgam; and, what is more, they have made results, in quantity, with their plastic, easy-working mass, which compare, most favorably, with any of the most exceptional successes of the "dry-mixed" amalgams.

Having by the addition of silver counterbalanced largely the shrinkage of the *plastic* mass, and having by the addition of gold still more controlled it and assured a better edge-strength and a better maintenance of color, it is *permitted* that we may make the mass of such consistency as will insure that facility of introduction and complete manipulation which has proven itself, in practice, to be adequate to the comfortable saving of dreadfully decayed teeth.

It is found, experimentally, that in proportion as alloys are *good*, so will they tolerate a large admixture of mercury; and, indeed, it may be stated much more strongly, that they will *require* a larger relative proportion of mercury for proper working.

The reason for this is evident; for, as the easily melted metal, *tin*, is replaced by those which are fused with difficulty, as silver, gold, and copper, it naturally requires more of the *amalgam making* component to effect the desired fusion.

I do not pretend to say that, even with these good alloys, the minimum of mercury, and the working in the powder form with heated instruments, would not produce a filling which might, theoretically, be better than one made with the larger proportion of mercury; but, for *practicality*, I must strongly urge the latter, as that form of amalgam which will be more conducive to such results as will be most *satisfactory* both to patients and operators.

Beside regarding the plastic mass as the best form in which to use amalgam, and the "mortar-mix" as better than the "hand-mix," from the greater cleanliness, facility, and *promptness* with which the amalgam is made, it is essential to the best result, with any amalgam, that the mix shall be accomplished in definite proportion with *one admixture*.

If a certain proportion of mercury and a certain proportion of filings mixed and rubbed together are known to make an amalgam which responds in a superior manner to the varied tests for excellence, it is by all means desirable that the two be placed together and made into amalgam in those definite proportions. If a certain definite result is accomplished in a given time by the rubbing together of these two — mercury and filings — at *one admixture*, the same definite result cannot be accomplished in the same given time by the gradual addition of *alternating* portions of first one ingredient and then the other. And yet this is the *usual* manner in which amalgam is made — filings or mercury are placed in either hand or mortar, and the other ingredient is added in such proportion as is *thought* right. If the mass is too plastic, and is more in quantity than is judged sufficient, the mercury is *finger* squeezed out, and with it indefinite proportions of metals, which destroys the harmony of mixture in alloy just in proportion to its original exactness.* If the mass is too plastic, and all is required, more filings are added. If it be then not sufficiently plastic, more mercury is added. Now all the time that this work is being done, *another work is doing* — the work which is in progress on the part of the ingredients; that work which eventuates in the *proper hardening* of the amalgam mass when it is properly incorporated.

The *overheating* by combining with too much mercury melts out certain of the ingredients in unknown quantities. If the alloy is a two-metal alloy, combined without regard to test results, of course comparatively little injury will ensue; the material, not being good, cannot be made very much less good; but in a fine alloy, compounded with great care and in such proportions as give excellent test results, the withdrawal of any proportion of any of its ingredients is, presumptively, detrimental, and the withdrawal of indefinite quantities of all its ingredients is, as I have said, injurious in degree just in proportion to its original excellence.

The *overcooling* by addition of too much filings has just the opposite effect; it chills the mercury in such wise as to prevent

*See Appendix, Sec. 4.

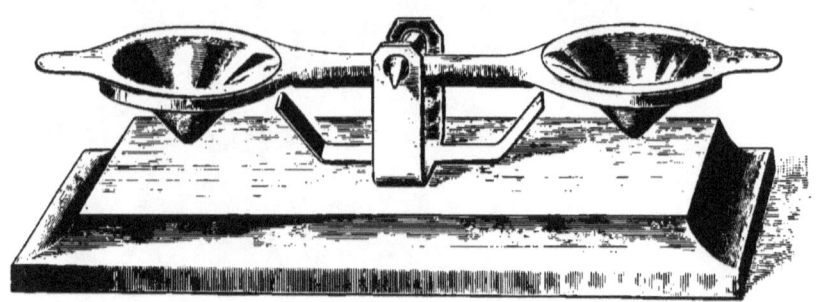

THE MAKING OF AMALGAM.

a given amount from doing its proper work of fusion. This is distinctly shown in the prompt manner in which an insufficiently plastic mass of amalgam will "take up" mercury. Old alloy, long filed, will take up by this unscientific mode of mixing as much mercury as new alloy, freshly filed, which is about one-fifth more than is required to make a good plastic amalgam. This at once transforms an excellent amalgam into a comparatively poor one.

Therefore, it is directed to prepare amalgam BY WEIGHT in such proportions as have been found upon trial to make the mass *just right for working in any given case.* This does not direct that the weighing shall be *by weights,* but by proportions; and it will be noticed, in the weighing of different alloys, that different proportions of mercury will be required for each *class.* As in the "quality test," this work will at once *grade* any alloy, subject to modifying influences — admixtures of cadmium, etc. — which have been indicated; and, governed by these, it will require for *first* or *lowest grade* of two metal alloys — *tin* and silver — about 37 to 39 *per cent.* of mercury; for *second* grade — *tin,* silver, and gold — about 41 to 43 *per cent.;* for *third* grade — *silver,* tin, and copper — about 46 to 48 *per cent.;* for *fourth* grade — *silver,* tin, copper, and gold, or *tin,* silver, gold, and zinc — about 48 to 50 *per cent.* of mercury.

This percentage of mercury is given as *average,* and may be diminished a little for perfectly easy work in very accessible cavities, but must be increased a little for that increase of plasticity which is demanded for acceptable working in the difficult and delicate manipulation inseparable from the filling of inaccessible cavities in frail and sensitive teeth.

In weighing, for the making of amalgam, the usual small brass scales of the jeweler is all that is required, though several devices especially for this purpose are for sale. The desired quantity of mercury for any given "make" should be placed in one scale, and the beam should then be *sufficiently depressed* upon the side of the alloy by pouring the requisite quantity of filings into the other scale. It will be understood, that for first grade alloys the beam will have to be notably depressed; for second and third grade alloys the depression will be less and less; for fourth — or highest — grade alloys the beam will deviate but slightly, if at all, from the horizontal. From the scales, the mercury and filings can be poured, successively, into the mortar; and this possibility should be provided for in any piece of weighing apparatus. If the beam is a fixture, the scale-plates should be movable, and in no case should the scale-plates or receptacles for the mercury and filings be secured to the beam, as this prevents emptying them in any other way than simultaneously. The scale-plates should be made of material which will not be affected by the mercury, as ivory, glass, porcelain, or brass.

This use of scales, instead of being troublesome, can hardly fail, upon trial, to be accepted as the easiest and only reliable way for promptly securing accurate proportions in the preparation of amalgam.

The proportionate quantities of ingredients having been obtained, they should be placed in a small *ground glass* mortar. For this use the porcelain mortar has been recommended and glass mortars condemned; but this has been from the fact that the glass mortars were used *smooth*, and in such it is almost impossible to make an amalgam; but if the glazed surface is *delicately* taken off the inside of the glass mortar, it will prove much superior to porcelain. In porcelain mortars, the face of the mortar discolors most markedly and disagreeably; the amalgam adheres in specks upon the sides, and it is impossible to remove it completely. In ground glass mortars, the face is easily kept nicely clean, and the amalgam mass is readily removed without much effort.†

The pestle should also be of glass, and the glaze should not be removed from any portion. It is also much better that the little knob on the small end of the pestle be broken off and the pestle be inserted into a wooden handle — hard wood — as this gives one more power for making the amalgam.*

Having the filings and mercury in the mortar, and having carefully removed any particles of amalgam which may be adherent to the pestle from previous makings, the filings should be *gradually* incorporated with the mercury. This is done by retaining the mercury in the centre of the bottom of the mortar, and by a circular motion, *occasionally reversed*,

---

*Glass pestles are now made which have proved even more satisfactory than those with wooden handles. †See Appendix, Sec. 3.

drawing in the filings little by little. This should be accomplished with sufficient deliberation, and yet with sufficient rapidity, a procedure which can alone be acquired by practice. The object gained by this process is, that in this way the percentage of mercury given will "take up" and make a *plastic* mass of the given proportion of filings, whereas, if the filings are at once, and without method, mixed with the mercury, it will become chilled, and the amalgam will be hard, dry, and crumbly.

The rubbing should be both decided and somewhat prolonged, until a smooth plasticity is imparted to the "make." The possibility of nicely accomplishing this result is the indication which governs the knowledge as to proper quantity of mercury. If this result is attained too easily, there is too much mercury; if it is not attainable, there is an insufficiency of mercury, and a little more should be added.

Amalgamation being, to a certain extent, complete, any amalgam adherent to the pestle should be scraped off into the mortar, using a knife or small spatula—of the set of instruments—for this purpose. The pestle is sometimes cleaned by rubbing it off with the finger, but the use of the spatula is much neater, and will, by practice, soon become habitual.

The mass is then gathered by the finger from the mortar into the palm of the hand, and is kneaded until it becomes a "button." This may seem a very simple performance, but it is one of the *best tests* for the *plastic-filler*. It is almost unexceptionable that, as the worker in amalgam takes his mass from the mortar and works it into a "button," so does he grade himself as skilful or unskilful in the working of the material, and so closely does he do this that an observing expert will place him to a nicety. It seems as though all his knowledge concentrates itself upon this little act, just as a singer manipulates his "telling note," and, inversely, ignorance and incapacity *will* crop out when one of the uninitiated attempts to handle a "button." It is the aim, then, by strong, energetic, decided effort, to knead partial amalgamation into homogeneous plasticity. The mass is gathered together in the palm of the hand by the forefinger and thumb of the other hand, and is then squeezed and smoothed into commencing homogeneity by a down stroke of the forefinger; then manipulated as before, and again squeezed with the down stroke. It is during this manipulation, oftentimes repeated, that the "crepitation" of an amalgam is heard. This was referred to in the article on "Tests" as "peculiarly pleasing to the initiated," for it is regarded as indicative of an excellent alloy. The short, sharp, decisive cry is something like that produced in the bending of a bar of tin, and with all the testings in which I have engaged, and even in the daily work of making amalgam for actual practice, it is invariably the case that, when it is perceived by one who knows its meaning, the recognition is immediately made manifest by facial expression of approval or by outspoken words of commendation.

The "button" having been made, it should be held in the palm of the non-operating hand by closing the fourth and little fingers upon it, thus leaving the thumb, fore-, and middle fingers free to aid in operating. This is at first by no means an easy thing for every one to do, but is acquired, without much difficulty, by practice, and is of decided importance, as it maintains to the amalgam a degree of plasticity, by warmth, which cannot be utilized if, as is frequently the case, the "button" is laid upon the operating-table, and there cut up into pieces. The amalgam is now ready for insertion.

ARTICLE X.

INSTRUMENTS FOR THE INSERTION OF AMALGAM FILLINGS.

WITH the idea that experience indicates the employment of *few* but *accurately adapted* instruments as conducive to expert manipulation and productive of superior results, I have gradually introduced, discarded, modified, and selected shapes, sizes, and number of instruments until, for daily routine work, I have for the last ten years had no necessity for change or addition; nor have I seen the possibility of any advantageous reduction in number.

Based upon these conclusions is a set of filling instruments

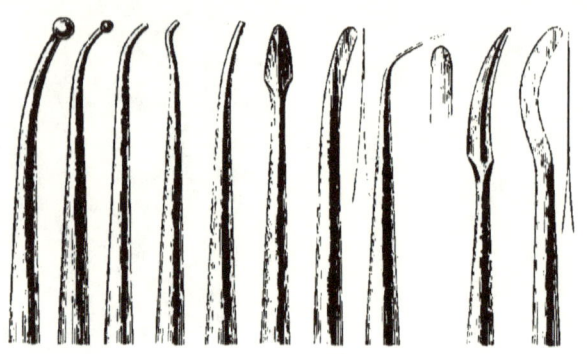

for amalgam, and, incidentally, for zinc-phosphate and gutta-percha stoppings, the patterns for which I have furnished for manufacture for the past ten years. The annexed illustrations, with descriptions for use, will, I think, afford all needed instructions to insure a satisfactory adaptation of means to ends.

It will be observed by practitioners who are generally conversant with instrument forms that I have endeavored to utilize familiar shapes rather than devise new instruments; and in this I have almost entirely succeeded. My reason for so doing is, that the habit of use which has been acquired in one direction may be made subservient in other work; thus the use of the ball-burnisher, having been acquired for the purpose of finishing gold fillings, may readily be made subservient to the packing of amalgam and zinc-phosphate, etc.

To the grouping and to the ideas of adaptation I have striven to give the impress of originality, and trust that in this I shall have afforded such aid as is desired by the inexperienced, and possibly may have been fortunate enough to give some hints of value even to those of large amalgam experience.

The set consists of twelve instruments, ten of which, Nos. 1, 2, 3, 4, 5, 6, 7, 8, 10, and 11, are especially adapted to amalgam work, and eight, Nos. 2, 3, 5, 6, 7, 8, 9, and 12, are subservient to incidental and special requirements in plastic filling.

It must be noted that the set is first divided into three classes —*round ends*, *flat ends*, and *trimmers* or *separators*. With these three types one is able to meet all indications. These classes are subdivided into

Round ends.—1. No. 1. Large ball-burnisher. This is used for *crushing* and *packing* the pieces of amalgam which are successively placed in position — either by thumb-pliers or amalgam-carriers — *in large cavities easy of access*.

2. No. 2. Medium-sized ball-burnisher. Used for the same purpose and in the same manner as No. 1 in *medium-sized cavities* easy of access, and in *inaccessible* cavities of *large size* for final packing after the pieces of amalgam have been preliminarily crushed by appropriate instruments,—carriers or flat ends,—also for tapping "wafers" (see Article XI.) and smoothing fillings upon the *articulating faces* of teeth,—especially of the lower teeth, — and upper second and third molars.

3. No. 3. Small-sized curved burnisher. This instrument is one of varied usefulness and of frequent adaptiveness. It crushes and packs the amalgam in a large proportion of cavities upon articulating, distal, buccal, and labial faces; and in cavities of all sizes above minimum it is in constant demand. It packs into under cuts, grooves, and corrugations; it crushes the outside pieces in large cavities — *especially buccal* — and packs them nicely, using its side; it smoothes the face of buccal, distal, mesial, and labial fillings; removes surplus material, and defines cervical edges of fillings extending below the gum; "wafers" buccal fillings better than any other form of instrument, and becomes, in amalgam work, an *indispensable*.

4. No. 4. Very small "goose-neck" burnisher. This little instrument is used exclusively in the filling of *very small* cavities. These are usually in the sulci of molars and bicuspids, or upon the necks of teeth, or in the mesial or distal faces of incisors and cuspids. Its peculiar form is eminently adapted to this work, as it is possible to insinuate it dexterously in spaces of very moderate size between teeth, and then so turn it as to compress and pack filling material in the bottom of cavities of a depth which is considerable in proportion to their size of orifice. This instrument is one of those which, though not of such frequent use as some others, is, nevertheless, the instrument for which there is no substitute when its services are required.

As I have spoken here of *very small* cavities in front teeth, and as it will be some years yet before the filling of such cavities — even in soft teeth — with plastics will become generally accepted as proper practice, I wish here to say that some of my *most decided successes* of the past twenty years have been made by this practice. I have scores of young patients whose elder brothers and sisters had been placed under the best gold work which our own and other large cities afforded, and whose front teeth — particularly laterals — had been filled and refilled with gold until pulp after pulp had died, and tooth after tooth had discolored,— and all this before they were twenty years of age, — whose teeth are of as soft, *and even softer* structure than those which had been so treated, and who are now from twenty to thirty years old, and who are *most markedly* reaping the benefits conferred by plastic fillings in "pin-head" cavities of front teeth, in that they rejoice in *handsome-looking teeth containing living pulps.* I have found it comparatively an easy task to care for such teeth *if the small cavities are filled first with plastics* — gutta-percha and "lined" cavities, with amalgam; for the work is easy and gentle, and thus the little patients are not demoralized. The fillings last as long, and usually much longer than do gold ones; and thus they do not require so frequent renewal. And in this way, when patients reach that time of life at which *they recognize* and *appreciate* the value of beautiful and entirely vital front teeth, *they have them in such shape as that they regard them as worth care and attention.*

I think there is *much* in this that is worthy the careful, thoughtful consideration of every dentist, and particularly of those who are now in the earlier years of practice. They will have the teeth of the *children of this generation* — TEETH BORN OF ARTIFICIAL DENTURES! — under their charge. It will be for them to make the first onset in antagonism to this wholesale loss of teeth, and my experience is, that "compatibility of filling material with tooth-bone" is the foundation for tooth salvation, and that *gold* is *eminently incompatible* with *soft* tooth-bone.

5. No. 5. Medium sized, slightly curved, round end. This instrument is used for crushing and packing in all perfectly accessible cavities of *medium* size, and is particularly adapted for "wafering" the faces of articulating fillings. This instrument is also the "large-sized probe" which is frequently mentioned in my pathological papers as proper for carrying all *thin fluid* medicaments, as hamamelis, oil of cloves, tincture of aconite, campho-phenique, etc., in the treating of teeth.

Flat ends. — 1. No. 6 is a short, double convex, spatula-shaped instrument. It crushes and packs the superficial pieces of amalgam in accessible fillings between teeth, where there is ample space, and particularly such wafers as are used for these fillings; it trims upon articulating faces in occasional cases, but is mainly used as a packer and smoother.

2. No. 7 is thin, almost flat, slightly curved, and is used, as is No. 6, in places where want of space will not permit of using a thicker instrument; it is also used as a trimmer in such cases as require No. 6 for a packer.

3. No. 8. This is a very useful instrument, and is used in inaccessible places between teeth; distal or mesial cavities; upper or lower teeth; cavities extending to and under the gum. It is particularly adapted to the *placing in position* of portions of quite plastic amalgam in very inaccessible cavities in both upper and lower teeth — distal faces — after which the pieces are to be packed with No. 2. In wafering such fillings as these, No. 8 will be found to work admirably; it is also useful as a smoother and shaper of fillings between teeth, where space will permit; also as a trimmer upon articulating faces, particularly of the lower teeth.

Trimmers or *Separators.*— 1. No. 9. This is a modification of the curved bistoury. It is more useful in zinc-phosphate work than in amalgam; but it is sometimes the case that the "quick-setters" harden too much for the easy removal of surplus material. When this is so, the alternating possibilities of the convex and concave knife edges of No. 9 will become agreeably apparent.

2. No. 10 is a modified Lord scaler of thin gauge; the point is rounded so as to prevent undue wounding of the gum in the work for which it is appropriated.

I have declared myself in favor of "contour filling," and although this conclusion may be unjust to those who "separate" between good, strong teeth, I must not be subjected to their criticism, from the fact that my work is *almost exclusively* pertaining to frail, soft teeth. These come to me largely requiring complete contouring, oftentimes with two-thirds and even four-fourths! of the crowns gone, and with but little masticating surface generally,— from scarcity of teeth; and I aim to make as close proximity of articulating faces — mesially and distally — as is possible.

These thin trimmers or separators are used for the purpose of cutting the thinnest spaces between such crowns, or between built crowns or fillings, and contiguous teeth. No. 10 is especially indicated in such work between incisors, cuspids, bicuspids, and occasionally even as far back as between first and second molars.

3. No. 11. This is also a modified Lord scaler, but it is so much modified that its suggester would hardly recognize it.

It, however, becomes well adapted for the work of cutting separations between molars, both above and below, by the making of the obtuse angle bend.

I would direct attention to the fact that the instruments of this set are arranged for "universal" working,— right or left, upper or lower; and it is claimed that much greater facility of instrumentation and celerity of operations are attained by the habit of constant work with few instruments, and the nonnecessity for the frequent change incident to "right and left" filling instruments.

As the complement to the dozen, I have suggested a spatula which, for shape, balance, size, weight, and temper, seems to me to meet requirements more exactly than any heretofore made. If I may judge from the expressions I have already received, I am correct in this, and I therefore counsel, more unhesitatingly, the use of an instrument which has so many rivals.*

In amalgam work, the spatula is never used for its legitimate purpose,— that of mixing the zinc plastics; but it will occasionally subserve excellently for smoothing fillings between incisors, particularly of the lower jaw, where space is limited, and where also, from the narrowing necks of the teeth, it is peculiar in its shape.

Besides the instruments for ordinary use, we have a number of ingenious appliances as aids in difficult cases. These almost invariably pertain to cavities in upper teeth, less or more inaccessible. The difficulty in such cases is the placing of the amalgam in position, and its subsequent crushing and packing, without great loss from dropping of the material. With the ordinary instruments this is sometimes impossible of accomplishment, and indeed it is, in some instances, exceedingly difficult to do even with the best appliances.

An instrument which is at once simple and effective is the double-end "carrier and plugger," devised by Dr. W. St. Geo. Elliott,— No. 1,— one end of which — the larger — is deeply serrated and filled with amalgam. This is allowed to harden, and, as pieces of fresh amalgam mass will adhere to hard amalgam, they are thus readily carried into position, and crushed with comparative ease. From the fact of this adhesion, the instrument is called the "*Loadstone* amalgam carrier."

---

* The *taper point* of the spatula is an *essential*, as this permits the accurate placing in position of the zinc-plastics for "lining." See p. 193.

The other end is serrated. I should prefer a smooth end, like a medium-sized ball-burnisher; but even as made, it is, I think a very desirable instrument.

Another device of the same gentleman is that which is illustrated as No. 2, and is at once ingenious and practical. Upon careful examination, the illustration will give a very good idea of the working of this instrument, although it is somewhat difficult to understand even by description. Upon the end of the shank is placed, diagonally, a cylindrical hammer-head; around this is a tube about one-fourth longer than the hammer-head — just the size illustrated. In this tube there is a slot of about one-third its length, and through this slot the neck of the instrument passes. By pressing, with the end of the middle finger, the downwardly inclined end of the investing tube, it is forced out beyond the hammer-head, and thus a receptacle for amalgam mass is made. By pressure on the mass, retaining the finger in place, a portion is forced into the receptacle. It is then presented to the cavity, when, by upward pressure upon the handle of the hammer, it is forced along through the slot, and the amalgam, ejected from the receptacle, is driven into the cavity; it is then to be packed with any appropriate filler. This is an excellent instrument, and is in harmony with the degree of plasticity of mass required for filling such cavities as it is intended shall be filled by its aid.

The only carriers which I have seen that rival, in the least, the one described, are those of Dr. H. S. Chase — No. 3 — and Dr. Thomas Fry — No. 4; both are ingenious and effective Of the other carriers illustrated,— Nos. 6 and 7,— they either do not do that which is required, or do it much less easily and efficiently than those more particularly referred to.*

Another form of carrier is that known as the "Amalgam Director" — No. 5. This is the invention of Dr. E. R. Mullett, and consists of a peculiarly shaped spoon, with a light and appropriate handle. It is a very neat instrument, and, with a little experience, will prove a complete preventive to loss of material in the filling of these inaccessible cavities.

It is as well adapted for carrying hard amalgam as it is for soft mixed mass, and for this reason is the only one at all use-

---

* I now regard the double-end spoon-shaped carrier, pattern of Dr. W. C Foulks, as the best I have ever used. See Plate, Fig. A.

A
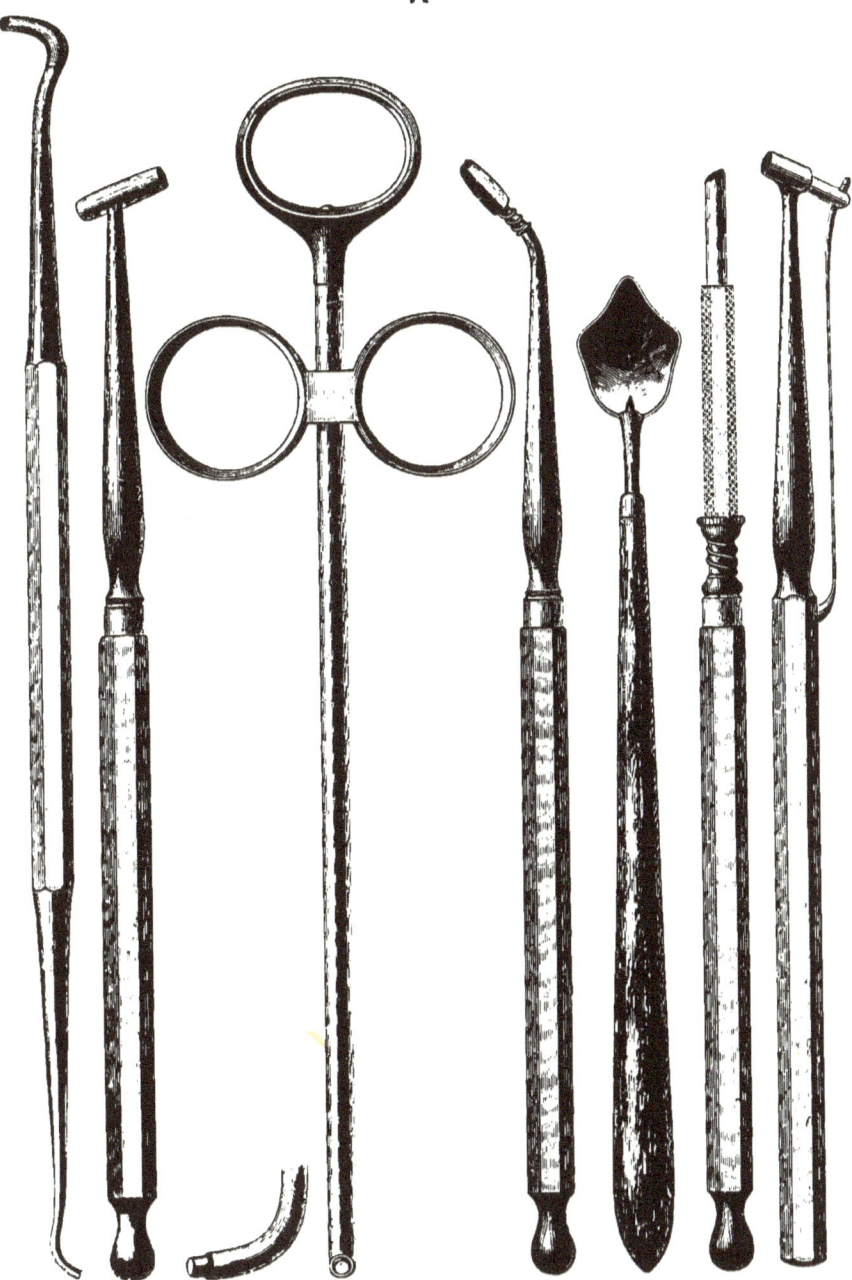

ful in wafering. It is particularly valuable, in "directing" amalgam into buccal cavities both superior and inferior.

It is not so good for carrying amalgam to cavities upon the *distal* faces of upper teeth as either the Chase or Elliott carriers, but it conduces to cleaner work, and is more frequently advantageous than any other carrier.

In this connection, I should not omit the information that, in addition to many appliances in furtherance, advantageously, of amalgam work, there are also quite a number which are truly curiosities in their way. Some are made with one end about equidistant in form between a toothpick and a sail-needle, and with the other end in the shape of a little cup so attached as to be incapable of subserving any purpose, even if there was any purpose to subserve; others have ends something like shoehorns, and handles like modern ice-picks; in short, the singularity of form illustrated by these instruments is only exceeded in degree by the absolute absence of possibility that *anything* in the way of amalgam working should ever be accomplished by any of them.

ARTICLE XI.

THE INSERTION OF AMALGAM FILLINGS.

AS a large proportion of the alloys which are at present in use will not make amalgams which can be graded even as "good" under testing, it is but reasonable to infer that but few operators have ever worked amalgam that would rank as "excellent" *upon its own merits*. This being the case, it must equally be presumed that the working of fine amalgam, differing as it does, *essentially*, from that made from all ordinary — 60 tin, 40 silver — alloys, will have to be acquired by nearly every dentist. Even those who have had considerable experience with such amalgams as are made from Townsend's, Walker's, Arrington's, etc., will find such decided difference in the working of Lawrence's, Hardman's, and "Contour," as will surprise them. And it should also be known that *their* working of these amalgams must not be regarded as forming any criterion as to the *possibilities* of amalgam work with such material, for, with observation, care, and practice they will soon learn that — as with gold — the most expert manipulator makes the finest results.

It is too generally believed that *an amalgam filling is an amalgam filling* no matter what it is made from, how it is made, or in what manner it is inserted; but this is hardly more correct than it would be to assert that *a gold filling is a gold filling*. It is true, that the range between extremes is not so great in amalgam work as in gold work, for the difference between a wretched gold filling and an admirable gold filling is so immense as to be incredible to any but proficients in such work; neither is the difference in results between poor amalgam filling and good amalgam filling in the least degree comparable with that which obtains between poor gold filling and good gold filling; for a good gold filling will save even a soft tooth for a certain length of time, while a poor gold filling may truthfully be regarded, from the first, as a very questionable occupant for a cavity in a frail tooth. On the other hand, the cases now count by thousands in which confessedly poor amalgam fillings have already done longer and better service in frail teeth than equally confessedly excellent gold fillings had done previously in the same teeth. Neither is the range between extremes in manipulative skill nearly so great in amalgam work as it is in gold work; but this I regard as one of the strongest points in favor of amalgam. It is a fallacy that the profession of dentistry bases its capability for rendering service to suffering humanity most largely upon the manipulative skill of its members. In by-gone times this may have been so; but the capability of dentistry *to-day* rests most largely — *and decidedly most largely* — upon the scientific attainments of its practitioners in the varied directions of anatomy, physiology, chemistry, metallurgy, pathology, and therapeutics. Skilful manipulation in both operative and prosthetic dentistry can never be underrated, for it leaves its impress, as does the finely-cut die upon the coin; but it can *never again be overrated*, as it has been in the past; it cannot, in future, grade the position of the dentist; it will ever be recognized as an essential, but it will be *subservient* to greater essentials.

With this marked change of estimate, *the insertion of amalgam fillings* has had, probably, more to do than any other one thing; no other plastic possesses the requisite characteristics for demonstrating so unmistakably that in dentistry, as in other professions, "*knowledge* is *power*." This arises from the fact that, though skill is required in the working of amalgam, and though skill does amply make its mark, it is nevertheless possible to do *tooth-saving work* with it *far more easily* than with gold. From this reason it is not needful that the energy of the student be so exhaustively expended in the direction of finger education; nor is the list of those who are capacitated for rendering truly creditable and compensating service, necessarily, nearly so circumscribed. This insures a *wider latitude of benefit to those who suffer*, and, inversely, it must redound to the *elevation of dentistry* upon the broad ground of *greatest honor to that which gives the greatest good to the greatest number*.

It seems proper to prelude the directions for the insertion of amalgam fillings with certain remarks discussive of the condition of cavities preparatory to their introduction. In order that this shall be complete, it must first be stated that "excellent" amalgam is at present divided into four kinds—"submarine," "usual, or contouring," "front tooth," and "facing." Much effort has already been expended for the production of a "universal" alloy; but, as yet, I have never seen any which made amalgam that met all indications. It appears to be as impossible to make such an alloy as it is to make a form of gold which will prove equally acceptable in all cases.

The old-fashioned *precipitated copper* and *silver coin* amalgams may be regarded as the typal "submarines," for although infrequently indicated to-day, they nevertheless give the basal idea for permanence in submarine work. It has been noticed that the *black* amalgam fillings done fifty years ago have saved the worthless shells in which they were introduced—possibly *by the finger* of some operator—merely as experiments, until, in some cases, every other tooth has passed away.

Analysis of these fillings shows them to be composed of either mercury and copper, or mercury, silver and copper—coin amalgam—and it is from this start-point, then, that we work for *tooth-conservation* as a distinguishing characteristic of amalgam.

It has also been noted that the most remarkable instances of long maintenance of integrity at the "vulnerable spot"— cervical edge — and that, too, under most disadvantageous circumstances — frail walls, soft structure, beneath gum, *under moisture* — have been attained by these *black* fillings. These, therefore, form the basis for amalgam intended for "submarine" work in difficult and inaccessible places.

I have stated that the thorough amalgamating of coin filings is a protracted and difficult piece of work; also that *tin* is a notable facilitator of fusion. It has therefore been deemed advisable to add this metal, in small quantity — from 30 to 35 per cent.— to alloy, and to substitute a portion of grain silver — about one-third — instead of using all coin. This makes a "submarine"— 60 silver, 35 tin, 5 copper;— the only positively determined alloy which practically meets the *average* of combined requirements better than coin amalgam. and an experience of fully twenty years gives to it remarkable "tooth-conservation" even under extremely unpropitious circumstances. It discolors reasonably and with desirable promptness, if exigencies demand it, but does not usually become as black as *pure copper* or *coin* amalgam, and does not impart to the tooth that greenish hue so distinctive of copper in quantity, uncontrolled.

Amalgam made from alloys of this nature may be successfully worked in cavities from which it is impossible to exclude moisture. It *can* be worked, and with reasonably good results, — the best, I think, excepting those of *coin* amalgam,— in cavities which are filled with moisture; but in such I prefer that the moisture shall be pure water, in which is dissolved sufficient carbonate of soda to render it slightly alkaline, rather than the fluids of the mouth, even though these be largely composed of the free flowing outpouring of the sublingual and parotid glands.

For this reason, I guard, as best as I can, with napkins, against flow of saliva; have the cavity prepared and filled with cotton; make ready, with decided plasticity, my "submarine;" fill my syringe partially with the alkaline solution, tepid; with draw the cotton; deluge my cavity, and quickly place in position and compactly insert my filling material.

The second kind of amalgam is that employed for "usual" or "contour" work — from smallest to greatest. All alloys based upon the "60 silver, 40 tin," modified by additions of gold and copper, would naturally come into this class. They work with nice plasticity; set well; shrink but little, if any; have sufficient edge-strength; hold their color well, and offer these varied desirable attributes in compensation for certain loss of "tooth-conservation" which, in *usual* work, is not so imperatively demanded.

In working this amalgam — uncombined with other plastics — it is undoubtedly true that just in proportion to the absolute dryness of the cavity, and to the maintenance of dryness during the thirty minutes required for "setting," other things being equal, will be the perfection and durability of the filling. In these cases, then, the plastic-filler uses the rubber dam when other appliances will not insure sufficient dryness.

It will be noted that this use of the dam takes from it all its horrors, and much that is no more than disagreeable, for there is *never* any need for passing rubber or ligatures below the gum, and, if a clamp is occasionally employed, its placing and necessary strength of grip are, neither of them, such as would prove inflictive, even to a child.

The formula which, as the result of the record of the past twelve years has been finally accepted, is the *ninth* of the original line of work done in 1876, 1877 and 1878 for the determination of a "contour" alloy — 63.3 silver, 33.3 tin, 3.3 gold. This makes an amalgam which is quick-setting, of excellent edge-strength, good color, and desirable plasticity.

This is the field for present work in connection with amalgam. It must certainly be recognized that, in the direction of submarine and ordinary work, the labors of the "New Departure Corps" have been such as to place the composition of future alloys upon a definite basis, which essentially differs from that almost universally adopted, and which decidedly sustains the direction in which had been developed the Lawrence and Hardman alloys. I think that this "contour" alloy offers the nearest approximate to the *average* of all requirements in usual or contour work that has yet been attained. The "quality test" is a *touchstone* which, I believe, will soon serve to weed out inferior and *injurious* alloys, and compel their replacement by those which are, at least, much better; and still the model alloy for contouring is not yet attained.

Work may, with advantage, be done in this direction — a work based upon those data which, from previous work, I have been enabled to present to the profession; a work which can be taken up by all investigators and experimenters at this point, with the assurance that the old landmarks have not been removed without long and careful consideration, and that the new ones have not been set up until they had been subjected to much thought, scientific scrutiny, and experimental observation.

The third kind of amalgam — "front tooth," as it is now known — is that used for filling soft front teeth which have been "disked out" or "separated" in such manner as to preclude the use of gutta-percha, and in which the cavities are either too small or are of such shape and position as to interfere with advantageous "lining."

These alloys of copper, silver, tin, gold, and zinc, possess the requisites of very good maintenance of color and of non-shrinkage, but are slow setters with *sufficient* edge-strength; and they very well fill the niche for which they are intended, and seem to complete the list of means which enable the plastic-filler to enter into such eminently successful competition with the gold-filler in the work of saving the front teeth.

As the result of an almost incredible amount of experimentation, running through a period of ten years, I had arrived at the conclusions that *gold* or *copper* were the metals upon which to base further work for the obtaining of alloys that should subserve *acceptably* in "front tooth" amalgams.

Always regarding silver as an *essential*, I argued that *gold* should control it up to the highest karat compatible with sufficiency of edge-strength — combining tin in small quantity as a flux — and admitting zinc for its well-known *maintenance* of color, while eschewing it on account of its supposed interference with edge-strength.

In this direction four years of work gave me, as most satisfactory, the *fifth* formula of the series—50 silver, 20 gold, 24 tin, 6 zinc.

Many fillings of amalgam made from this alloy have already done several years of acceptable service; but again, in too many instances, the presence of the silver, in quantity, was eventually made known by the gradual discoloration of the fillings.

The "working" of this amalgam was very pleasant, the "setting" was reasonably quick, the "edge-strength" was "sufficient," and for some time hopes were engendered which the results of years too largely dispelled.

In the *copper* direction work has been upon a different basis. Silver was recognized as a noted blackener! Copper was recognized as, if possible, excelling in the same result! And yet, it will be seen in the first edition of this book, page 50, that I was early impressed with the *whiteness* of the *alloys of copper and tin*. Four years of work upon this line gave me as most acceptable for a "front-tooth" copper alloy—38 copper, 37 silver, 25 tin. This alloy is very hard to mix, requiring *several times* the usual amount of rubbing before any sign of amalgamation is given; but when once commenced progress is satisfactory, though *very thorough* mixing is absolutely necessary. The introduction of the amalgam is satisfactory, but "wafering" as *thorough* as the mixing is again *essential*. No ordinary squeezing, with ordinary pliers, will at all suffice, but *thorough wafering*, with *properly-made wafering pliers*, *must be done*.

This amalgam is a wonderful color-keeper—it even gives to teeth with frail, thin cavity walls a color whiter than natural—but the filling is not quite satisfactory as to edge-strength, and when worked but little less perfectly than has been directed it sometimes "sweats." This result may not be looked for under a week or two or three, dependent upon degree of thoroughness of manipulation; but, if not sufficiently thorough, tiny globules of mercury come to the surface and soften the face of the filling. When thoroughly worked I have never seen this result. *Eventual discoloration* was accepted as the concomitant of the *gold* line of work, and *deficiency of edge-strength* was accepted as that of the *copper* line of work.

The basal formula for copper work was 70 copper, 15 silver, 15 tin; and when it was noted that admixture of the proportions of the "best" of the *gold* line with those of the "basal" of the *copper* line produced an alloy singularly analagous to the *whitest* of all, and yet diminished somewhat the tin and substituted a large per cent. of gold, the experiment of such an admixture was naturally tried. Given then, by mechanical mix, the formula of 35 copper, 33 silver, 19 tin, 10 gold, 3 zinc, we have that result which is the most acceptable of any at present appertaining to the line of work upon "front-tooth" alloys.

The fourth kind of amalgam "facing"—55 tin, 40 silver, 5 zinc—is one which is, *possibly*, least of all liable to discoloration. Its attributes are entirely in the direction of subserviency to its *three special* requirements, viz., the "facing" of bicuspids and molars (see pages 181, 182), the filling of "tap-holes" in positions liable to attrition, and the *proper securing* of arsenical applications in difficult and inaccessible positions. Moderate in its setting, deficient in density, wanting in edge-strength, it yet subserves all these purposes very nicely, enabling the plastic-filler in many cases to attain better results than can be secured with any other material. As a "facer" or a "tap-filler" its advantages can readily be appreciated; but as a cover for arsenical applications in difficult positions it may be well for me to say that in my experience I have found no other material in the least degree comparable to it for *ease of adaptation*, absolute *exemption from leakage*, maintenance of integrity during period of application, and final *facility of removal*. For this purpose *alone* "facing" amalgam would be invaluable.*

Copper Amalgam.

It seems to me that any *discussion* of this material is certainly a step backward into the *dark* ages—not that it has not *value*, for that has always been conceded to it; not that it has not *a record for tooth salvation*, for its only competitor in this regard is *coin amalgam*. This fact, I believe, is universally conceded. Knowledge in relation to its original methods of preparation and manipulation is "as old as the hills," and it therefore results that there is nothing to *discuss* pertaining to the material, and that there is but one thing known of it to-day that has not

---

* The union of "facing" amalgam with zinc-phosphate in "combination" filling, and the employment of "facing" amalgam for the securing of fire-gilded porcelain fillings, are utilizations of this amalgam which are to be specially noted.

been known ever since the days of the fathers. A quarter of a century ago it was *known* that it would save some of the poorest, softest teeth for a quarter of a century. To-day it is *known* that it has saved some of these very teeth for *half a century!* And this is the *only knowledge* that has been gained regarding it during the past twenty-five years.

But other considerations decidedly do pertain to the "craze"—as it has most appropriately been termed—which has been alike the astonishment, the disgust, the amusement and the regret of many "lookers on" in dentistry during the past three years.

Astonishment that those "eminents," who ten years ago could tolerate nothing but *yellow* and who fairly despised *black*, should now be found *in droves* espousing *black* in such degree as even to assert that like or dislike in color is purely educational! and that a beautiful *jet* is a beautiful thing!

Disgust at witnessing the rampant repetition of the old stereotyped claims of the ordinary alloy makers from all the numerous manufacturers of copper amalgam.

Disgust at reading the preposterous and manifestly untrue assertions that each is "best"—how can *each* be "best?"—that all are "chemically pure!"—as though chemical purity is a *needed advance;* for if the old, so-called "Sullivan's amalgam" *was* "chemically pure" there is *no advance;* and if old "Sullivan's" was *not* chemically pure, as it has saved teeth for *fifty years*, it would seem to argue that chemical purity was not such an all-important need!

Disgust at such unkind, unwarrantable assertions as that "American touch" had more improved the methods of manufacture and *quality* of copper amalgam in three years than had *other touch* in half a century. And in what did all this wonderful improvement consist? In the substitution of *ampere dynamo,* run by "expert electricians," for simple plates of iron or zinc, and of *hydraulic presses* for expressing the mercury in place of a piece of chamois skin and a pair of pliers!

It seems fortunate that American claims for "progressive" work usually rest upon foundations far different from such as these, as the only progress thus far *demonstrated* is increased trouble and vastly increased expense without one particle of proof that anything *better* is obtained as the resultant, for it will take *fifty years* to *prove* that any of the "chemically pure," "scientifically compounded," "made by a new process," "hydraulically compressed," "electrolytically precipitated," "pronounced BEST by all who have used it" copper amalgams, are in any degree *better* than the old, so-called "Sullivan's."

Disgust at the periodic "methods of manufacture" which, being given in the guise of instruction, are so intricate in their verbosity and so confusingly prolific in their minutiæ, as evidently to be intended only to impress the hearers with the exceeding complexity and delicate accuracy of the work of compounding copper amalgam!

Amusement, mingled with chagrin, at the gradually increasing rush of the unsophisticated who have demanded by the *many thousands of ounces* the despised, insulted, condemned material of their grandfather's, paying *two and a half dollars per ounce* for that which their grandfathers *made for themselves* for about *seventy cents per pound!*

Regret that the upheaval should show so conclusively the unfortunate instability of the foundations for the practice of that dentistry which could possibly permit the presentation of such a picture.

Regret at the manifest ignorance which such a picture incontestably confesses; but perhaps *not so much regret* at this tardy confession of the correctness of the estimate of the "plastic workers" as to the value of even this, *the lesser reliable* of the two old "submarines," for, as was said at Niagara in 1878 by one who is yet a "notable," "If my life depended upon the durability of my filling in a soft tooth, I would fill it with a *silver-coin* amalgam."

And so would I.

The Making of Copper Amalgam.

To one ounce (troy) of pulverized sulphate of copper add eight fluid ounces of warm water. This dissolves the salt promptly. A piece of thick bar iron ($\frac{1}{4}$ inch thick by $1\frac{1}{2}$ or 2 inches wide, and from 8 to 10 inches long, is about right for a small jar or glass) should be *brightened before using* by a "pickle" bath (sulphuric acid, 1 part; water, 3 parts) of sufficient length of duration—from half an hour to an hour. Into the clear blue solution immerse the plate of bright iron. The formation of solution of sulphate of iron and the precipitation of copper, in fine pulp, commences immediately. This process continues for several hours, and the copper pulp should occasionally be scraped off the iron plate into the solution by means of a piece of stick. When no more copper precipitates, the iron plate should be taken out, cleaned bright by *thorough brush-washing*, and dried. It is thus left ready for future use.

The supernatent solution (of copperas) is of a yellowish-green color, and when it is reasonably cleared by the settling of the precipitate it should be carefully poured off. Sufficient "pickle" (three or four ounces) is now poured upon the precipitate, and this is frequently stirred with a glass rod. This is for the purpose of cleaning and purifying the copper pulp, and should be done at intervals for several hours until no bubbles rise from agitation. A much greater length of time seems rather beneficial and not at all prejudicial, and I therefore leave the pulp in pickle, sometimes for several days, occasionally stirring it until I desire to finish the work. When sufficiently pickled the acid-water is poured off, and the soft, pulpy residuum is put into a glass or wedgewood mortar, to which is added one ounce of mercury. The whole is then rubbed up with a pestle, and prompt and complete amalgamation ensues. After the amalgam is made it is washed repeatedly, and sometimes a *trifle* of bicarbonate of soda is added to one of the waters. I have found thorough washing to be, experimentally, sufficient without the soda.

The mass is now squeezed *gently* through "chamois" skin, and the half-squeezed mass is thoroughly triturated in the mortar. It is then squeezed hard—as for wafering—and is again triturated, divided into small portions, and is either packed into moulds or squeezed into wafers and allowed to harden. This takes from twelve to twenty-four hours, and the amalgam is then ready for use.

To utilize it for filling purposes it is placed upon a small, flat and somewhat thick spatula, and is held over the flame of a spirit lamp until the mercury "beads" upon the surface; then with the warm spatula it is crushed into a semi-plastic powder,* which under the instrumentation incidental to the introduction of

---

* After crushing, the powder is rubbed to a paste in a mortar.

the filling becomes quite plastic, sometimes permitting increased wafering.

Copper amalgam "sets" with exceeding deliberation, but eventually becomes *very* hard; has excellent edge-strength; does not spheroid; its shrinkage is but little; does not, except in soft teeth, discolor tooth tissue, though it *blackens absolutely* itself.

Copper amalgam has a notable record as a tooth-conserver, although its record is by no means without blemish; on the contrary, the failures of this filling material have been multitudinous and quite varied in their characteristics. I have thought that this was due to improper or careless making of the amalgam; overheating, or "burning," as it is called, of the material prior to its insertion; to imperfect manipulation in its introduction, thus leaving it in a granulated or powdery condition; or to the addition of mercury in excess when preparing it for final use in order to obtain what is truly an *unnecessary plasticity*. Be this as it may, causes have existed which have so depreciated the record of copper amalgam as to place it decidedly below coin amalgam as a *reliable* material for filling poor teeth.

I have given what is known as the "iron method" for making copper amalgam, because I have always regarded it as the more trustworthy; the other method, known as the "zinc," having been accredited, though perhaps falsely, with a large proportion of unreliable copper amalgam. In my own experiments I have noted a diversity of results from the "zinc precipitation," which I have never noted from that by iron, and it is for this reason that I have long since adopted exclusively the "iron method." This will be found in practice very simple, very easy, very economical—*less than ten cents an ounce*—and more than all, very reliable and uniform in its produced amalgam.

Note.—The zinc method having always had its advocates, I therefore suggest for this the use of a plate of rolled zinc instead of the thin sheets usually employed; this obviates completely the presence of flakes of zinc amongst the precipitated copper. And further, that the solution of sulphate of zinc is *clear*, thus affording proof of the complete precipitation of the copper. Also, that the pickling process be long-continued, as the rising of bubbles from stirring the precipitate will continue for two or three days. With all this care the amalgam thus made sets more slowly, has not so good edge-strength, and is liable to a diversity of results not pertaining to the "iron method."

With both methods I have tried "cleansing" or "purifying" the precipitate with everything that I have heard suggested, from salt to sugar, from nitric acid to carbonate of soda, from chalk almost to charcoal! and, with the exception of the essential *sulphuric acid* and *possibly* for its neutralization a *little soda*, I have failed to recognize any advantage accruing from their use.

I would mention a method of utilizing copper amalgam which, although it complicates the work at the time of operating, yet so perfectly eliminates all four of the causes mentioned as liable to conduce to questionable results, that I have thought it worthy of repeated trial. For this purpose the pickled copper precipitate is thoroughly washed and dried, and is kept in a closestoppered bottle in the fine, soft resultant powder. For use, powder, say ten grains, is placed in mortar; forty grains of mercury is added; the powder is moistened with water and a few drops of sulphuric acid, and the amalgam is made, washed thoroughly, squeezed gently, rubbed again, squeezed as for wafering, broken up and rubbed again, and is then introduced into the cavity of decay and wafered as usual.

In concluding this subject I feel it my duty to say that I have written upon copper amalgam to supply that which seems to be a *present* want, and not because I attach any special importance to it — for in my practice, *confined exclusively to soft teeth*, I have had no other than occasional experimental use for it for nearly thirty years.

Coin Amalgam.

During the earlier years of my amalgam work, and even so late as when the conclusions of the "New Departure Corps" were presented to dentistry, so many practitioners were perfectly conversant with "coin amalgam," either from utilization or from hearsay, that I did not deem it essential that it should be other than casually referred to; but during the last decade so many of the elders have passed away, and so many recruits have joined the ranks who have no knowledge of the attributes, record, or even of the past existence of this filling material, that it seems essential that this "*dernier ressort*" should have place, were it only that a good word may be spoken for it at this late day after the extended and thorough reviling it has had for the past half century.

I have always felt that had it not been for "coin amalgam"

the varied list of filling materials, now known under the comprehensive name of "amalgams," would never have had existence, for it is the *proven*, wonderful tooth-conserving power of this "silver paste" that *alone* has enabled it to "hold its own" against the fearful assaults of the "ancients." It has not the sharp and strong edge of copper amalgam; it turns black and discolors teeth dreadfully, both soft and *hard;* its "setting" is rather deliberate; its spheroiding and its concomitant crevicing are decidedly not satisfactory; *but against all this is its unbroken record of tooth-salvation.* In many a cavity where everything else, even copper amalgam, has failed, it has done faithful duty *persistently.*

The Making of Coin Amalgam.

Any regular silver coin—90 silver, 10 copper—should be cut into filings by using a "dead smooth" file. The filings should then be sifted through a *very fine* sieve, magnetized, and blown as in the usual preparation of alloys. For filling take mercury and filings—about 55 mercury to 45 filings; scales *decidedly* down on mercury side—rub into paste in mortar; if amalgamation is not prompt and complete add a globule of mercury to make a smooth paste; squeeze through chamois; again triturate thoroughly in mortar; squeeze as for wafering; introduce filling as usual; take off soft surplus, and wafer. The peculiar, disagreeable, granular feel of the material under the instruments, and especially during the first finishing, together with the "crumbly" condition of the surface of the filling, is the accepted thing with "coin amalgam" work, and is excessive just in proportion to coarseness of grain of the filings, and it is to obviate this as much as possible that the "dead smooth" file is used. After its introduction the filling should remain two or three hours and then be smoothed *gently* with a burnisher. I have found the small-sized curved burnisher, No. 3 of the set, the most universally adapted to this work. The filling should then be allowed to *harden completely;* and although coin amalgam "sets" even sooner than copper amalgam, I nevertheless give two or three days for this purpose, at which time, or at any convenient season thereafter, it should be smoothed, stoned and burnished in a manner worthy of so excellent a material.

Amalgam mass, having been made ready for insertion, is held in the palm of the non-operating hand by the closed fourth and little fingers. The palm being turned up and the fingers raised from the button, a portion of amalgam is cut off by the thumb-pliers, rolled into an oblong spheroid or spindle-shaped pellet upon the palm of the non-operating hand by the forefinger of the operating hand, and is then taken by them and — the button being again covered — placed in the cavity. If this be in a lower tooth, the piece will naturally lie in position; but if it be an upper tooth, the piece will have to be retained in place by one of the free fingers or thumb of the hand containing the button.

The piece should now be crushed by one of the round-end or flat-end pluggers, and thus secured in position; *it must then be tapped with light blows* from an appropriate instrument until it is placed in accurate apposition with the walls of the cavity. It has long been the practice to *rub* amalgam into contact with cavity-walls, and to rub the various pieces consecutively into union with the amalgam already introduced, but the filling of the lower part of two small glass vials will easily convince any one of the decided superiority of "tapping" over "rubbing." A piece of paper should be pasted around that portion of each vial which is to be filled. This prevents watching the progress of the filling and renders the vial, in this respect, more like unto a tooth. Then fill one vial *very thoroughly and very carefully*, taking extra pains, by the rubbing and burnishing method. Then fill the other, with only ordinary care, by tapping the pieces into position and into homogeneity the one with the other. Scrape off the paper, and it is quite probable that the most skeptical will be entirely convinced as to which is the better method.

It has been stated that amalgam should be malleted, either by hand or automatic mallet. I regard this as not only unnecessary, but really objectionable. It is not needful that the blows should be in any degree *forcible*, but, on the contrary, it is better that they should be such as would more properly come under the signification of "taps." The consistency of the amalgam should be such as will permit of perfect adaptation of filling to cavity-wall by tapping with light blows.

In the same manner as is cut off and introduced the first

piece, so should the succeeding pieces be cut off and introduced, each piece being united to its predecessors by *tapping;* without this precaution, the unions are imperfect, lines of demarcation between pieces are clearly apparent, and leakage is inevitable.

The mass should work quite plastic—generally becoming more and more so as the filling progresses—until the cavity, if it be of ordinary size, is completely filled, when a "last piece" should be enveloped in a fold of chamois skin and *squeezed hard*, and the wafer thus made be laid on the filling. By crushing this wafer into powdery pieces and tapping these into homogeneity with the softer mass, the setting will be hastened, and the edge-strength, density, and whiteness of the filling are much increased.

This process is called "wafering," and will be discussed in the article on "Technicals of Plastic Filling." I have said that if the cavity be of ordinary size, the process of wafering should be used after the cavity is completely filled; this is so because the size of the filling is not so great as to prevent a sufficient effect upon the whole mass from one wafering; but if the cavity is very large, or if it is inaccessible, it is better that two or three mixings be made, in small quantity, and that each mix be hardened and set by its own wafer. This will be found particularly advantageous in "building" crowns, or in making large reparations, as contouring amalgam may be so worked by this treatment as to set almost as fast as it is built on, and to become sufficiently hardened in an hour or two to subserve the purposes of mastication or of clasping.

After the filling is wafered, it should be shaped by trimming, or, if not too hard, it may be smoothed into shape by a piece of soft pine stick. For this purpose, I have pieces of white pine—the softest, and of perfectly straight grain—prepared in slips of six inches long, three-eighths wide, and three-sixteenths thick; the ends of some are cut to a small round for articulating faces, and the ends of others are cut to a delicate chisel shape for smoothing buccal surfaces and for finishing between teeth. Smaller pieces are adapted to a Cogswell, or other wood carrier, and are thus used in inaccessible places.

After the filling is smoothed, it should be allowed to harden for ten or fifteen minutes; it should then be burnished, and smoothed again with the stick; then allowed to harden for another ten minutes, when it should be finally burnished and *finished white* with the piece of stick, using, if necessary, a little very finely levigated pumice. In smoothing for final finishing, the stick should be passed over the face of the filling *up* or *down* and not across the filling; this leaves the lines of finish so that the light will strike upon presenting parallel faces, and thus gives to the filling a much whiter appearance.

In such amalgams as are controlled by copper and gold, this face will frequently be very well maintained, and I therefore think it advisable to permit the opportunity for their doing so, but if the filling should discolor, it can at any subsequent time be filed or burred off, smoothed and burnished. Such amalgam fillings will, almost universally, retain the brightness of this burnished face.

It has been sometimes directed to burnish the edges of amalgam fillings after they have assumed a partial hardness. I can only regard this practice as beneficial with the very poorest amalgams, such as shrink notably, have but little edge-strength, and crevice markedly; but I should equally regard it as positively detrimental to a filling of good amalgam, and detrimental just in proportion to the excellence of the amalgam. A filling made of excellent amalgam should have a clean, sharp edge, of great strength, which should never be broken by burnishing, but, on the contrary, should be allowed to harden thoroughly, and depend upon its own physical characteristics for its integrity.

ARTICLE XII.

GENERAL CONSIDERATIONS PERTAINING TO AMALGAM.

IN taking up this subject under the three divisions of: 1. Local effects; 2. Systemic effects; 3. Possibilities, I cannot help premising that, from my standpoint of observation and experience, the outlook is something so novel to the typal dentist of the golden age as to appear absurd, and, even more than this, wild and visionary; but I am quite sure that to the younger men, those who have inherited the *leavings* of the past half century's efforts at saving (?) teeth, those who have to grapple with the daily task of overcoming such apparent impossibilities as would not even be attempted by the most enthusiastic workers in gold more frequently than once a week, and, more than this, to antagonize and conquer presentations that would be utterly hopeless in usual practice, or else continue in the wholesale extraction of the present, and help to flood the nations with artificial dentures, my views of amalgam will, at least, seem worthy of perusal and of thought, and that it will be recognized by them that what has been done so satisfactorily as to have engendered such opinions, might well be tried in vastly more extended fashion, and with the determination to give to the experimentation the *benefit of every doubt.*

I feel strongly the truth of the position that amalgam has been so long decried, and is so very meritorious, that I may be pardoned for urging its claim with evident partisanship. I feel that in doing this I cannot do the thousandth part of the great wrong to humanity which has been done by those who have, for many years, so successfully debased it to a low position. I feel that it has been *used* by the very men that have *abused* it, when they could use nothing else, and, that when it has responded to their call most grandly, they have ranked it as *secondary* the more decidedly, from very fear that its *facts* might supplant their *theories.*

This is strong placing; but it is none too strong; for, by this doctrine of subordination of this valuable material, by this teaching that in its requirement of little skill, and in its glorious attribute of easy working, *it lowered its manipulator and degraded dentistry,* thousands of dentists have been restrained from using it when it would have been the very best thing known for the requirements, and have been stimulated to the employment of gold in cases where it has constantly exhibited the most signal lines of failure even when worked by the few who were dignified by the ranking of "eminently skilful."

Tens of thousands of cavities have been thus filled at fearful expenditure of time, suffering, and expense, which *ought to have been* quickly, comfortably, and economically filled with amalgam.

Hundreds of thousands of teeth have been lost, years before their time, from no other cause than that, in using gold, impossibilities were attempted in defiance of incompatibility of material, and in despite of the insurmountable physical characteristics which precluded its working; while, with amalgam, these years of comfort could have been bestowed through its nearer approach to compatibility, its peculiar tooth-saving property of forming soluble sulphides, and the physical characteristics which permit its easy working under most disadvantageous circumstances, and in most inaccessible cavities.

Through the dogmas of the superiority of gold as a filling material and the elevating of its working to the position of "standard for excellence," and of the inferiority of amalgam as a filling material and the degrading of its working to the position of "standard for incapacity," *millions* of teeth have been *needlessly sacrificed*, and their places filled with those cunning devices of quartz, feldspar, and kaolin which are so appropriately designated — *substitutes*.

My voice is given in antagonism to such teaching and to such practice with full confidence, born of years of severe testing, that a change cannot, by any possibility, produce any more discreditable results; and with a firm hope that it may give comfort, health, and *teeth* to those whose fathers and mothers have lost all these at the hands of *first-class accepted dentistry*.

1. *Local effects.*—As I have intimated, in previous articles, the local effects of amalgam are not altogether what we would desire; but, as I have also stated, most emphatically, they are in great degree eminently desirable. It therefore behooves us to discuss, clearly and dispassionately, this important matter in its perfectly tangible bearings.

I desire that it shall be understood that I regard it as needless now to discuss, at any length, the peculiarities pertaining to ordinary amalgams; for it is a principal object with me to give such information regarding "plastics" as shall force all ordinary materials from the market, unless they are offered *as such*.

While this arrangement will exclude the vast majority of amalgam alloys at present manufactured and used, it will yet permit the retention of some that are worthy of limited continuance, at least until they are replaced by various makes which shall possess generally good characteristics.

But it will also permit the retention of some which are given as possessing the very objectionable distinguishing characteristic of decided tendency to *discoloration*. This peculiarity, then, must be accepted, *with its concomitant tooth discoloration*, at its precise valuation; it must be viewed as unfortunate that we have not yet attained any non-discoloring amalgam which we can positively declare to be as *tooth-saving* as those which freely discolor; but it must also be taken as partial compensation for this, that we have succeeded in establishing the remarkable controlling of copper, zinc and gold over this discoloration, and that such amalgams are notably creditable as tooth-savers.

It must be recognized that if an amalgam which discolors—as, for instance, made from alloy of silver 60, tin 35, copper 5—is chosen as a filling material, its *local effects* are regarded as such as overbalance in good that which is admitted as bad. It works plastic against frail walls; it works as a submarine; it sets with desirable celerity; its shrinkage is infinitesimal; its edge-strength is good. All these are, indisputably, desirable qualities; but it permits the formation of sulphides, and in so doing produces two results—the filling becomes darkened and presents an unsightly appearance, and the tooth assumes a darkish hue, sometimes even blue-black, which is also eminently disagreeable to view. But, again, this very process is the *saving of the tooth*, and it therefore becomes simply a question of *utility* vs. *beauty*, which, it seems to me, could in every instance very readily be promptly and satisfactorily settled between patient and operator by a plain presentation of facts.

Another local effect is the occasional *induction of galvanic electricity by contact with other metal*. This is most usually developed when such contact is made by touching fillings with pins, needles, metallic toothpicks or forks,—either steel or silver,—and the results vary in degree from merely making a peculiar taste, galvanic, through the entire range of sensation from disagreeable to intensely painful—a shock.

When this occurs, the filling should be removed and an insulating medium—either zinc-sulphate, varnish, oxy-chloride, temporary stopping, water-proof court-plaster, or any equivalent to these — should be so placed as to cut off the sentient conductivity, and thus preclude such liability.

This effect is quite liable to be produced from contact with the clasps or plate of artificial dentures — gold — and is sometimes, though rarely, difficult of prevention. It is in some cases readily obviated by so filing the plate as to preclude its touching the filling during the insertion or removal of the work and during mastication.

If the unpleasant effect is due to contact of clasp with filling — either in clasp teeth or adjoining teeth — the means for prevention are more complex, less positive, and require more careful consideration. Regarding *difference in potential* and *brightness of metallic surface* as the combining circumstances which, under peculiar condition of oral fluids, produce this result, the first effort should be the removal of brightness.

This is done by oxidizing the clasp by heat, and by "facing" the bright-surfaced amalgam with a thin layer of "submarine," mixing a small pellet of this kind of amalgam and rubbing it on those faces of the filling which are touched by the clasp, and then removing it to original contour as the thinnest possible film is sufficient; or, by coating the inside of the clasp with a mercurial covering by rubbing upon it a small portion of amalgam, which if done carefully, so as to impinge upon no part of the work except the inner portion of the clasp, will not spread, neither will it enter into the substance of the clasp sufficiently to be detrimental; or, by changing the location of the clasp and securing to another tooth; or, by using silver for a clasp, leaving it darkened from the blow-pipe; or, by changing, if possible, the filling material to some other, though less reliable and less durable plastic; or, *as a final resort*, if "clasping" *is a necessity* or a *very great comfort*, the devitalization of the pulp of the clasp-tooth.

This latter consideration is one which it is supposed will be subject to that recognition of contingencies — probabilities and possibilities pertaining to pulpless teeth — which, with the knowledge of the present day, are so accurately reduced to debatable propositions. In mouths where this galvano-electric effect is very decided, it will occasionally be noted that the contact of small pieces of amalgam with gold fillings will produce a like effect. Thus, the dropping of *débris* during the introduction of amalgam fillings, or the making of contact during the rinsing of the particles from the mouth, will occasion more or less decided *starting* on the part of the patient, and will require the desired explanation. In filling under rubber-dam, this difficulty would, of course, not occur; but in napkin filling, the napkin should be so arranged as to secure, if possible, the catching of all dropping filling material.

A third effect is like unto this second, but is occasioned from the fact that *a filling of different metal, having marked difference of potentiality (usually gold), is so inserted—disconnected, not touching—into the same tooth or an adjoining tooth as to be occasionally connected by tongue, lip, or cheek connection.* Whenever this connection is made—usually during mastication—a shock of sufficient severity to cause pain and starting is felt. This is cured, either by thoroughly amalgamating the face of the gold filling, or, if in one tooth, by cutting away between the two fillings and *making them join*, using either amalgam or gold for this purpose, as is indicated either from considerations of appearance or ease of manipulation; if it arises from fillings in two teeth, one of the two fillings may be removed and replaced with a like material to the remaining filling. I should, of course, advocate the removal of the gold filling and its replacement either with tin, gutta-percha, or amalgam, according to circumstances, as I should suppose the considerations which prompted the introduction of amalgam in one of the cavities would outweigh the objections which might pertain to the use of one of the other materials mentioned. It will be noted that effects like these could only occur in wholly vital teeth.

A fourth effect is *a species of metallic salivation* due to action through the "gustatory" reflected through the "nutrient" of the salivary glands. This is a very rare occurrence, though I have met with a few cases of it. I have always classed it as physiological rather than pathological action, and regarded it as similar to the effects occasionally produced through the optic and auditory nerves at the sight or mention of certain kinds of food. This effect is sometimes of but short duration—a few days—and, as the faces of fillings lose their brightness, it gradually passes away. Having noticed this, I act upon the hint, and in cases where it is persistent, and in which the faces of fillings maintain good color markedly, I remove them and replace with amalgam which discolors, carefully lining the cavities so as to prevent tooth discoloration.

A fifth effect is that of *giving rise to a bad taste;* this is even more rare than the preceding one, and does not occur more frequently than once in from five hundred to eight hundred mouths. In every instance, *except one*, which I have seen of this condition, the teeth were of that kind in which gold had been tried by various operators of acknowledged skill, with anything but satisfactory results, and in the one exceptional case I heard that a most extensive and elaborate line of gold work was finally resorted to. In these cases I have removed the amalgam and resorted to gutta-percha, oxy-chloride of zinc, and zinc-phosphate fillings with the requisite renewals.

A sixth effect is an irritation of the fauces, throat, and larynx, which, while it is not *primarily* dependent upon the presence of amalgam, is nevertheless much less amenable to treatment, while numbers of large amalgam fillings are in the mouth. In an amalgam practice of twenty-five years, and with an experience in more than five thousand mouths, I have met with but two cases of this kind. In both these cases serious bronchial trouble was previously constantly existent, but was decidedly less controllable under amalgam irritation than without it.

In both these cases I removed the amalgam and refilled with gutta-percha and oxy-chloride, and, more recently, with zinc-phosphate, and in both instances with relief to the patient.

It is more than probable that the vast majority of practitioners would pass an entire lifetime without meeting any such cases as these I have last referred to, but it is well to know that idiosyncrasies exist in which such things can occur, in order that one may be prepared to meet them should they happen to present themselves.

As for the other local effects, which are frequently charged to amalgam, such as pulp-devitalization, periodontitis, alveolar abscess, exostosis, and necrosis, I must say, most decidedly, that I have never met with an instance in which any of these conditions existed with amalgam fillings as concomitants, and in which I could not think that it would probably have occurred, under similar circumstances, in connection with any other metallic filling material.

2. *Systemic Effects.*— I have already spoken of the discussions which I asked from the Pennsylvania Association of Dental Surgeons almost thirty years ago. It may be supposed that, as I had been for six years directing close attention to all such matters as could be pretty well settled in that space of time, I would have especial interest in the views of my fellow-members upon a subject of such grave moment, and of such difficulty in establishing data *by one observer*, as that of the systemic effects of amalgam fillings.

I had already conversed with many of the elders of dentistry, and had found that, *without exception*, those who had had experience, either limited or extended, were alike positive in the denial of any such systemic effects as could be attributable to *mercurials*, while I had also found that the few who "believed" in this possibility, were also, without exception, either avowedly reiterating the views of those in whom they had confidence, but without having any personal knowledge in relation to it, or were openly hostile to its use, without ever having investigated its merits or demerits to the extent of even *one* filling.

In referring to the proceedings of the meeting of April, 1861, reported in "Cosmos," May, 1861, I find that I expressed thus early in my work the opinion that, while it was "barely possible" an idiosyncrasy could exist in which amalgam, *per se*, might be injurious, it was, nevertheless, a thing which I had never seen.

I quote briefly to show the collateral basis which was, upon that occasion, given me, and which I regarded as substantiating my own opinion sufficiently for all practical purposes.

Dr. C. Newlin Pierce, who opened the discussion, said, in regard to amalgam ptyalism, that it was a thing of "so rare an occurrence, that he believed the profession had never heard of but *one* practitioner who *thought* that that result was produced by amalgam."

Prof. T. L. Buckingham said "he had never seen a case of salivation from its use, and had doubts about its ever having produced ptyalism;" that mercurial effects were "influences produced through the general system; but he did not think amalgam fillings would produce these effects."

Dr. J. H. McQuillen said that "in an experience of fourteen years he could not recall a single instance of necrosis of the jaws, ptyalism, etc., of which others assert that they have seen so many;" and that while he recognized the fact of idiosyncrasies in which the smallest quantity of certain medicinal agencies is followed by untoward results, and would not, therefore, offer his negative testimony as positive proof, yet "his own experience had made him look upon those who assert that they have seen so many cases with considerable doubt as to the value of their judgment or opinions as reliable diagnosticians."

Dr. C. P. Fitch said, "in regard to its toxical or injurious effects upon the system, he was inclined to question, if not wholly doubt, any such influence, and concurred in the views advanced by Dr. McQuillen, that he had yet to see the first case of alveolar abscess, ptyalism, etc., due to the presence of mercury in the amalgam."

Dr. J. M. McGrath testified for himself and for his father, who had had an amalgam experience of ten or fifteen years, that as yet they "had never seen any bad effects resulting, such as had been ascribed to its use by many practitioners."

Thus it was that I was fortified by the combined testimony of gentlemen whom I esteemed as conscientious observers, and for whom I had much regard both socially and professionally.

It now remains for me to add the testimony of almost thirty years more of increasingly acute scrutiny, with the assertion that during all my amalgam experience I have *never seen one case* of mercurial ptyalism, mercurial periostitis, mercurial necrosis, or of the slightest symptom which could reasonably be ascribed to mercurial action. I have had cases of asserted mercurialization by the score brought to me. I have treated them experimentally with chlorate of potassium to demonstrate its utter impotency, and have then *cured every case* without the use of any anti-mercurials, and have left the teeth refilled with amalgam. If anything more convincing than this is required, I have it not to offer. And yet the cry of mercurial ptyalism still continues. It is repeated not only by medical men, for whom we must have the needed leniency, but it is stated boldly and persistently by dentists; by men who have, at least, local reputation, and are thereby capable of doing harm. But I wish it to be noted that no man of *record*, either scientific, literary, or professional — beyond the making of a good gold filling — can be found to-day who will commit himself by the assertion that, from observation and experience, he believes amalgam fillings are liable to produce mercurial ptyalism, and are, therefore, unfit to use in efforts for saving teeth.

3. *Possibilities.*—With this theme I am filled with such thankfulness at the knowledge of these, and with such gratitude that my professional career has been in the pathway opened to me by them, that I long to tell, *in one word*, of the possibilities of the various amalgams; it seems to me as though there was but one word which could express it, and that is *fulness*.

Of all the other filling materials, there is no one which covers a tithe of the ground in dental demands that is covered by amalgam *when used with a knowledge of its possibilities*. Far more than this, I most unhesitatingly assert that, in a practice which shall save every tooth and root that offers which can be saved advantageously, amalgam will enable an operator to produce more satisfactory results — grading these from an *average* of the varied stand-points of beauty, permanency, comfort, economy, and utility — than *all the other filling materials combined*.

It may well be regarded that I would not make any such statement as this without much consideration; and it may well be admitted that I have had an experience which has given me ample data upon which to base an opinion; and it is, therefore, upon the combined strength of reflection, experiment, and observation that I have ventured the position.

The *possibilities* of amalgam; it is an easy thing to tell of its possibilities, but its limits — for it has limits — are not so readily defined. It can take care of almost anything in dentistry; and it really seems as though its possibilities were more and more demonstrable, just in proportion as the result desired seems more and more *impossible of accomplishment*. The things which are easy to do with gold are easier done with amalgam. The things which are hard to do with gold are easily done with amalgam. The things which are exceedingly difficult to do with gold are not difficult to do with amalgam. The things which are almost impossible to do with gold are not difficult to do with amalgam. The things, by the dozen, which are *impossible* to do with gold are not very difficult to do with amalgam. And to do well, too; to do in such wise as that they shall be eminently useful, durable, comfortable, and satisfactory; to do in such wise as that the recipients shall, after years of trial, point to them as proofs to others that they, too, can enjoy such benefits.

And this is the material that for fifty years has been stigmatized as *base;* frowned upon as *low;* tabooed as unworthy the notice of skilful men; disgraced as the token of incapacity and quackery. Is it not time that this thing should be ended? Is it not time that its serious investigation should be undertaken by the coming men of dentistry?

Ordinary Cavities.—It has, for so long a period, been regarded as the proof of incompetency that an operator should even *think* of filling ordinary cavities of decay with anything but gold, that I deem it of exceeding importance that I should present this subdivision of my subject with that precision and decision to which I esteem it entitled.

It has been a systematic thing with me to do so for many years, for I was early impressed with the fact that *this* fallacy was probably, of all others, the *most pernicious.*

Who would think of allowing a small break in a dyke to become an enormous crevasse before trying means for its arrestation which had been demonstrated to be equal to extraordinary emergencies.

Who would think of allowing a small fire to increase in size until it should *demand* the services of a steam fire-engine, if one had the engine in readiness to extinguish the flames in their incipiency?

Who but an old school dentist would *order* that a small cavity in a soft, young tooth should be filled and *refilled* with gold, until, as filling after filling *dropped out,* the decay had progressed to such extent as to render it impossible that it could be filled with *anything but amalgam?*

And more than this, who but an old school dentist would

make *obedience to this order* the *test* for *capacity* and for "*respectability*" as a dentist? It is too late for the premises in these questions to be disputed; such practice is not only *advocated*, but it is *ordered*. Such *results* are not problematical; they are the *usual, almost invariable* results of such practice.

And deviation from this *rule* has always brought contempt upon the man who deviated, *that he might save teeth*, from all the members who were regularly recognized as "most respectable." This has been the tone of the vast majority of practitioners; it has been the tone of the society discussions; it has been the tone of the lecture stands; it has been the tone of the journals, both small and great—and thus it has been *made* the tone of the Profession of Dentistry.

It is a blessed thing for the patients that a few men *have deviated* from this practice, and in despite of deviation have had the strength to swim; for the most of those who have done so, have been forced under incontinently.

It is a blessed thing for the salvation of teeth, and for the consequent health and comfort of this rising generation, that the appeals from those who have thus "*departed*" from these old time landmarks, are arousing the members to a sense of the need for inquiry, investigation, and experimentation.

I trust, and firmly believe, that it will prove a blessed thing for dentists and for dentistry. I therefore teach that the *ordinary* pin-head cavities are those in which the work of saving teeth *is to be commenced;* that it may with propriety be inferred that a material which will do well in emergent cases will also do well in the control of trifling lesions; that a material which is easy of manipulation will be less demoralizing in the using, to the young sufferer, than would be a material requiring great skill and persevering labor on the part of the operator and commensurate quiet endurance upon the part of the patient; that a material which will average *more durable* work than gold, in difficult cases, may properly be accredited with the capability for making more durable work than gold, as its "average," in easy cases; that thus it is but reasonable to deduce that a marked increase of successful effort would result from the employment of amalgam in place of gold in such ordinary cavities as are so situated as to require a filling material possessing the physical characteristic of ample resistance to attrition during mastication.

It was in furtherance of this view that I spoke as I did upon the occasion of the presentation of the "New Departure" principles and practice, saying. "Front teeth, *par excellence*, are filled and refilled with gold, as the *best* that can be done, until pulp after pulp dies, tooth after tooth becomes discolored and crumbles away, root after root is extracted, and plate after plate is inserted. This is *stereotyped practice*, and I defy contradiction of the statement." "*I think* that a liberal allowance of gutta-percha and amalgam fillings in these very teeth, while yet the cavities are *only pin-head cavities*, would be *a step in advance of this.*"

"This is *still talk; it makes quiet;* but it is what I have come to-night to say."

Large Cavities.— It would seem to require no argument to show that in proportion as cavities increased in size, so would the demand increase for such service as had proven effective for the retardation of the loss of tooth-tissue in teeth largely affected by caries. It is nevertheless true that positions already begin to multiply which render the employment of amalgam in these cases additionally advantageous. With every grade of cavity enlargement, the labor of introducing a gold filling becomes greater in quite accurate ratio, while a notable difference in size of cavity makes but little, if any, increased labor attendant upon the introduction of amalgam. The expense of gold filling is also quite in consonance with the size of the filling — relativity of position being conceded — while this is but very little changed in the use of amalgam, whether the cavity be "ordinary" or "large." The *conductivity* of a filling is to be considered as detrimental in exact proportion to pulp approach, and, in this regard, amalgam, with its comparatively low conducting power, gains in superiority over gold with the loss of every successive stratum of dentine. Thus, on the varied score of tooth-conservation, saving of labor and of time, saving of expense and saving of pulps, amalgam presents its claims of superiority to gold for the filling of *large cavities* of decay.

Enormous Cavities.—To such patients as have had experience in the filling of enormous cavities, as demonstrated by the two schools of gold and plastic dentistry, any discussion of the methods and results seems purely ridiculous. I cannot place it better than it was done by one of my patients who had *enjoyed* more than a decade of years of each kind of practice after he had arrived at an appreciative time of life. He said, "If one lived five or six thousand years, and it had been positively demonstrated that it was a *great deal better* to fill teeth with gold, I *might think* of returning to that kind of practice, *but for this paltry seventy years, give me plastics!*"

This is the view from the patients' side of the question; and is it to be supposed that their comfort, their time, their health, their strength, and their teeth are of more importance to their dentists than to themselves? And yet, *this* is the dental view, and so the teachings continue that *it is better* to fill "enormous cavities" with gold.

It is conceded that it is exceedingly difficult to do, but, if well done, it is a *proof of superior skill;* it is conceded that the expenditure of time is vastly greater with gold than with amalgam, but it is taught that the difference in result *is proportionately better;* it is conceded that the expense is threefold, fourfold, tenfold, that of amalgam, but that with this *it is compensating, and eminently so;* it is conceded that in this work the drain upon the vital force, of both patient and operator, is something fearful to contemplate, but it is said that the *beauty* and *utility* of that which is produced *is worth it all.*

Now, I ask close examination of all this; I ask long and careful experimentation in regard to this; I ask that it shall be done with *knowledge* as its basis, throwing aside that easy soother of conscience, that salve of professional pride, that facile cover for ignorance which has been so frequently, and with such complacent suavity, referred to as "judgment." I ask that there shall be no more of this perversion of language. I would have it admitted that without *knowledge* there can be no "judgment;" and I would have it felt, and earnestly felt, that in this matter all should recognize the power for good which would ensue from the combination of *knowledge and judgment.*

I have worked long and hard to demonstrate the fact that

the weight of evidence is in favor of the filling of enormous cavities with amalgam instead of gold. I have taken dentures by the hundred which had been abandoned, by both patients and operators, upon the score of utter worthlessness, upon which months of time, hundreds of dollars, and mountains of agony had been expended, with no other return than repeated, signal failure, and with plastics, and *mainly* amalgam, have kept them, SATISFACTORILY, in working order, year after year.

That which has been done can surely be more easily done again. With all the guiding experience of the past, with all the improvements in composition, preparation, adaptation, and manipulation, it will be found that but little inquiry will be needed to prove, *first,* that, as a mass, enormous gold fillings are not *compensating*, especially to patients, and *second,* that a properly composed, properly prepared, and properly introduced amalgam, will prove, in truth, an advantageous " royal *metallic* succedaneum."

Entire Crowns.—It is quite frequently the case that caries continues progressing in despite of all efforts, local and constitutional, operative and medicinal, until the crowns of teeth are either completely destroyed or are rendered so frail as to be broken off during mastication. It is also very frequently noted that, although decay still progresses until the dentinal portion of the roots is almost gone, the cemental portion maintains its integrity in a remarkable manner, and thus, remains of roots are in position for fifteen or twenty years after their crowns have been lost.

It is these facts which have led to the practice of "utilizing roots," as it is called, for the purpose of "pivoting," "grafting," or "building on" crowns.

Various methods and very ingenious devices have been suggested for the securing of natural crowns and crowns of porcelain, amalgam, and gold; and for this *securing,* amalgam has already been extensively and very advantageously employed, and bids fair to almost entirely supplant every other material. It is probable that large and elaborate "crownings" will be done with gold, for, perhaps, many years yet, but it is not likely that such work will continue to be permitted, much less be demanded, by patients, as the evident *practicality* of plastics has even now resulted in their being very largely preferred.

The *first* use for amalgam in cases requiring entire crowns, is the "building up" of the crown with this material. This is done by one of three methods and by the use of either one, two, or three kinds of alloy.

If the decay of the crown has progressed so slowly as to permit the gradual covering of the pulp by deposition of secondary dentine, thus leaving a crownless root containing a well-covered vital pulp, the face of the root is prepared with especial regard given to the strength and solidity of the periphery and to the leaving of all living — though softened — dentine over the pulp. Retaining slots are then made by drilling, with spear-pointed drill, lines of two or more drill-pits at different selected places, and then connecting the pits by using a fissure drill or small oval burr. These should be filled with an amalgam made from "contour" alloy — approximate, silver 63, tin 33, gold 4 — for its qualities of non-shrinkage, quick-setting, and eminent edge-strength. The fillings should be made a little more than full, and rounded out, that surface for adhesion may be afforded to the tooth-conserving amalgam — made from submarine alloy — with which a base for the crown should be built out to the edge, covering the entire face of the root. The crown should then be built on with contour amalgam, which, by alternate layers of plastic mix and wafers, can be carried up, shaped, hardened, and finally finished at one sitting, and which will set sufficiently well in an hour to preclude any liability to accident if guarded with even reasonable care.

If the crown is upon a bicuspid root, or upon that of a first molar, which shows either during speaking or laughing, the buccal portion of the crown should be cut out as for "facing" — see Technicals — and the concavity be filled with facing amalgam. It will be seen that by this method the varied requirements of strong anchorage, excellent root conservation, rapid and strong crown formation, and beauty in presentation are all satisfactorily met. I have made some very serviceable and nice-looking operations of this kind upon cuspid roots.

If decay has progressed quickly, and a pulp exposure, devitalization and extirpation has ensued, the operation of crown building is much simplified as abundant retention is most readily obtained. In these cases the great advantage of amalgam becomes apparent in matters of vital importance connected with the duration of possible tooth maintenance, for the two great questions of irritation from dental manipulation, and easy possibility of arranging for relief from probable future trouble are most perfectly solved in the using of this filling material. The canals having been filled with glycerine, into which is passed taper-twisted canal dressings of cotton wool* dipped in fluid cosmoline, mixed with engenol or carbolized camphor, then covering the floor of the *pulp cavity* with temporary stopping, the filling is shaped into a hemisphere. Around this, in undercut grooves, is anchored the amalgam filling, which is then built up into shape and allowed to harden for fifteen or twenty minutes, or until it is enough set to permit of easy cutting without danger of breaking. An entrance is now made, by careful, gentle drilling or cutting through the filling into the temporary stopping. This entrance, if upon an articulating surface, is then almost entirely filled with gutta-percha, over which is placed a covering of amalgam. If it is upon any surface not exposed to attrition, it is, of course, completely filled with gutta-percha. If the relief route is closed with amalgam, it should always be done with "facing" amalgam, as this will leave a distinct demarcation indicative of position of "tap."

The third method of attaching amalgam crowns is by pins or wire-loop; for this, either platinum or soft iron wire may be used. I have attached a large number of crowns by each of these two kinds of wire, and I can see no advantage which platinum possesses over iron except that it is more easily bent. I have therefore gradually settled into the habit of using iron wire in cases where the pins were to be left straight and were required to be strong; and platinum where they were looped or bent, and were likely to require more accurate bending after being secured in position.

For either pins or wire-loop, one end should be rivet-headed, and in some cases of straight pins it is requisite that they should be double-headers; for the making of these I use a slide screw-plate, into the screw-hole of which the pin is

---
* The cotton wool should *not* be "absorbent cotton."

placed, and, when clamped by the slide, very quickly double-headed with a riveting-hammer.

Drill-holes, of the size of the rivet-head, having been made, and spheroided with a round or oval burr, the pin or loop is placed in position, and secured with amalgam. This is allowed to set, and these then form a firm reliance for the retention of the crown.

I wish here to speak of the advantage of this method of securing pins over that of screwing them into the dentine. For more than fifty years, the principle of screw attachment has been offered for the purpose of fastening crowns upon roots, but, so far as I have seen, the results show that it is not a desirable method. The efforts during mastication tend to produce an imperceptible loosening, which, trifling as it is, increases quite rapidly after being once established. Soon the motion becomes observable, and then, almost immediately, the screw-thread, acting as a file, cuts the pin loose, and the crown drops off.

Just the opposite result will ensue from the rivet-head and filling securing, for these have been proven to hold until the wire has been broken, or wrenched out by fracture of the root, or undermined by the progress of recurring decay.

The *second* use for amalgam, in cases requiring entire crowns, is for the purpose of attaching gold crowns. These are made by bending a ribbon of pure gold, cut from plate made of gold-foil scraps or old gold fillings, and soldering it as a ring with an approximate fit for the face of the root to be crowned. The anchorage is then obtained, and a filling of submarine alloy is made which extends a short distance out from the gum. When this is a little hardened, the gold crown is pressed into position, cutting through the amalgam, and thus forming a joint *at* the neck of the root, *not* by slipping over it and banding it, but by *resting upon it*, which joint is then made perfect by tapping the still workable amalgam accurately into apposition with the upper edge of the encircling ring of gold. Contour amalgam is now employed for filling the entire crown, and thus it is secured upon the root. One might naturally suppose that the mercury from the amalgam would permeate and discolor the gold, but it does not do this, and, from this fact, enables us to utilize this facile and durable method of fastening such crowns. This style of work formed my "intermediate" between the gold crowns which I formerly made with foi. and the amalgam or amalgam and porcelain crowns which I now use almost exclusively.

The *third* use for amalgam, in cases requiring entire crowns, is for the replacing of natural crowns which have been broken off. This is sometimes a very gratifying operation, both to patient and practitioner, for it is usually alike pleasing and unexpected to the patient to have such restoration made, and it produces a result which, for perfection of appearance, cannot be equalled by any other process.

The crown, possessing as it does all the special peculiarities which pertain to that individual location, if replaced, gives to the work a harmony with surroundings which cannot be otherwise attained; and it is for this reason that it is not only warrantable, but eminently desirable that the attempt should be made, especially for lady patients, and especially again for the conservation of facial identity or great personal beauty.

For this purpose the crown should be so drilled as to obtain the necessary mechanical hold for its part of the securing filling. The root should be drilled out deeply, and bell-shaped at the entrance, so as to give broad, strong retaining surface for the amalgam at the union between crown and root. The *labial* portion of the crown — it is supposed that this operation would not be deemed advisable for any tooth posterior to a cuspid — should be lined either with a very thin lining of oxy-chloride of zinc — in which there is little or no strength — or a somewhat thicker lining of "front tooth," which may subserve the purpose of increasing the strength of attachment: the object of which is the prevention of discoloration, for neither are strong enough to rely upon for support.

The root should now be nearly filled, and the tooth having been placed accurately in position, and being held either by the finger and thumb, or retained positively by gutta-percha, plaster of Paris, oxy-chloride of zinc, or zinc-phosphate supports, mesially, labially, and distally, should now be secured by contour amalgam. This should be allowed to set for at least a day, when the supports may be carefully removed. Such teeth should not be bitten upon for a day or two, if it is possible to avoid it, for it is very essential to success that the amalgam should be thoroughly set before it is subjected to the slightest strain. For increased strength, a small platinum-pin may be secured in the root and permitted to pass into or through the crown, as in a Bonwill pivot.

The *fourth* use for amalgam in these cases is for the securing of natural or porcelain crowns which have been fitted to crownless roots. From what has been said relative to the replacing of crowns which have been broken off, it will require but few words to present this modification of that operation; but it will be seen, upon slight presentation, that modifying considerations do exist. A crown which had been broken off would, under almost all ordinary circumstances, have been subject to extensive decay prior to fracture; while a selected crown — either natural or porcelain — would be perfect in contour. The one, the porcelain, would be so drilled and prepared internally, and at its neck, as to require but trifling adaptation by grinding, and subsequent securing by contour amalgam; the other, the natural, would require drilling and internal shaping of cavity to correspond with the prepared cavity of an artificial tooth; and it should then be "lined" to prevent discoloration.

Pivoting Teeth. — In this work, as in the whole line of operative dentistry, amalgam has wrought most radical changes; it seems to have supplanted the entire range of round and square box-work with split-pin — gold — pivots; it has largely taken the place of foil for securing these tubes and boxes, even if they are used. It has only gutta-percha as a rival for the securing of metallic-pin pivots; and it *welcomes* the aid of gutta-percha, for it recognizes in it a most notable plastic. It has essentially modified every phase of pivoting, and has rendered that operation, which was formerly either disgusting — wooden pin in plain root — or very tedious, painful, and expensive — backstayed-plate tooth built in with gold foil — a painless, comparatively inexpensive, and cleanly piece of work; equally beautiful in appearance; equally durable, and requiring but an hour or two for its doing.

Its first great change was the supplanting of the increase of size of wooden pivots, as the roots became more and more decayed, by the filling of the enlarged pivot-hole with amalgam, and the drilling of such sized hole as permitted a return to the original size of wooden pin. This was a great step in advance, as by it many a root was saved for many a year of usefulness, and even retained until further modifications resulted in the re-pivoting of teeth, in a superior manner, upon roots which had already supported pivot teeth for more than a dozen years, and for which fears as to their possible long continuance had been entertained many years before, as pin after pin had been introduced with biennial regularity.

The day of wooden pins is slowly passing away, and yet this use of amalgam will probably continue for a long time to come, and will be resorted to, as it now is, for foundations for first wooden-pin pivotings instead of permitting the roots to become almost ruined before lining them.

But the marked influence of amalgam over pivoting is most strikingly exemplified in the operations as variously proposed by Dr. Bonwill, Dr. Gates, and myself.*

Bonwill Pivot. — Select pivot-tooth, — Bonwill's; prepare root, and fit crown; fill root with "usual" amalgam — soft make — and force entrance for platinum-pin in amalgam with a pointed instrument. Grasp pin — which should be of right length, tri-flattened, and barbed — with forceps, and press it forcibly home. Tamp amalgam, to be sure that it well secures pin. This pin may equally well be rivet-headed and filled into the root. Replace crown to see that it fits. It is advantageous to bend pin so that it rests against palatal side of hole in crown, as thus the crown is retained nicely in place. Fill base of crown through to palatal side with amalgam — plastic — and with one finger over palatal hole press firmly into position by twisting, and afterwards tapping, by wooden pin, with mallet. The fingers will be best if the amalgam is not too stiff; then "wafer" to harden the amalgam. After it is thoroughly hard, smooth off the exposed portion of filling and the end of the platinum-pin if protruding.

I have set quite a number of these "pivots," and have reason to be well satisfied with the firmness, cleanliness, and appearance of the results, and with the reasonable facility with which the operation is performed.

* See Appendix, Sec. 6.

Although it is possible to make a Bonwill Pivot operation with almost any plastic-filling material, it is yet deemed preferable by the inventor that it be done with amalgam, and with this belief, which my experience amply corroborates, I have used no other material for many years. I occasionally use a tube in place of a pin, which I think adds greatly to the strength and value of the work.

For an accurate description of the Bonwill operation, and for its complete illustration, reference can be made to the "Cosmos" for August, 1880, p. 410, and for June, 1882, p. 304.

The Gates Pivot.—Such radical changes have been made in this "pivot," and such ingenious instruments have been devised for the accurate performance of the work, that I deem it better to refer my readers, for full information in regard to this operation, to the very explicit and admirably illustrated article in the "Cosmos" for March, 1882, p. 119.

Flagg's Ring Pivot.—This method of pivoting is so easily employed in all ordinary cases, and is capable of adaptation in so many difficult cases, that I have thought it advisable to describe it. Select ordinary plate-tooth; fit it to root, and bevel cervical portion of tooth as in "Flagg Pivot." Make small ring, of platinum wire, for straight-pin tooth, or semicircle for cross-pin tooth, and solder it to the pins above or below, as indicated—cross-pin tooth preferable. Take occlusion and cut off pin as required. Fill platinum pin in root.* Place tooth in position and bend ring, which should be of sufficient size to touch lingual edge of root, letting the ring slip over the pin. Secure tooth, with "adhesive wax," to adjoining teeth, and fill around pin and ring with Contour Amalgam,—soft make. The advantages of this method are found to consist in the almost limitless variety from which to select tooth; the latitude, which the ring allows, for placing the tooth in any desired relation with the root or with adjoining teeth, and the ease with which apparently insurmountable difficulties are overcome. It is usually advisable to allow the adhesive wax to remain, for a day or so, until complete hardness of amalgam is attained. For this purpose the amalgam should "set" for an hour, and the wax be then melted and smoothed into contour with the adjoining teeth by instrument No. 7, quite hot, as by this means a firm and neat

---

* For roots which may require "venting" use platinum tube for pin.

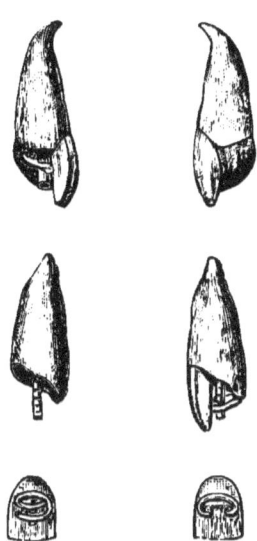

adhesion of wax is effected, and good temporary support obtained.

In cases of close occlusion, where strength of crown is impaired by necessary removal of material, I have found the "ring pivot" easy of adaptation and *very strong*.

Flagg Pivot.—Select plate-tooth, fit it to root, and bevel it from near the pin — cervical — or pins, if cross-pins, to the labio-cervical edge. Solder a platinum-pin to it as a backstay and pivot combined, leaving it rough or grooved on both sides of pin for a retaining hold to the finishing palatal amalgam.

Fill the root, which is treated, prepared for strong maintenance of filling, and "bell-muzzled," or "open-mouthed," with a good *usual* or *contour* alloy — quick-setter; non-shrinker; good edge-strength. I prefer to give this a day to harden thoroughly, but in case of need it may, with care, be worked in an hour or two.

Into the root-filling, drill a hole rather larger than the platinum-pin, as near to the *palatal* portion of the filling as possible, and directed, slantwise, to the apical centre of root-filling; then "fissure-drill" the hole towards the labial portion of the filling, trying the tooth until it sets just right, with the pivot-pin pressing hard against the labial side of the now *oval* pivot-hole. *By this method the tooth is accurately placed in position, and easily held firmly in place, while the pin is secured by filling the pivot-hole with amalgam.*

Let this harden for half an hour, and then add amalgam, in contour, to the root-filling and palatal face of the porcelain tooth. It is at this point of the operation that the need for "bevelling" the cervical portion of the tooth is demonstrated; for, by this bevel, one is enabled to make, by filling, a perfectly tight joint at the labio-cervical junction of tooth with root, and also to secure a strength of amalgam equal to the entire surface of root-filling.

This makes a strong, cleanly, and satisfactory operation.

"*Guarding*" *or repairing gold fillings, and refilling cavities from which gold fillings have been lost, in teeth which still contain gold fillings.*

We have now reached a singularly vexed question in dentistry, viz., the use of two metals in one tooth, or, possibly more dubious, the use of two metals in one cavity. In the discussion of this question, I shall regard amalgams as metallic compounds, each of which is possessed of that definite potential, conductivity, resistance to tarnish, etc., which makes of it, *practically*, a metal, and thus renders fillings of tin and gold, and amalgam and gold, equally *bi*-metallic, *dentally*. There is scarcely any point which has been more frequently spoken to in discussion than this, and probably no one upon which *more theory* and *less practice* has been expended in antagonism.

From the fact, which has been alluded to, that galvanic action can be excited by fillings of different metals in close proximity, and is variously demonstrated by metallic taste, pain, shock, and even spark, it has been largely concluded that the placing of such must necessarily be detrimental.

Reference to discussions upon this point will show that while only vague ideas of how this combination could work injury were enunciated, positive belief that it would do so was unhesitatingly and forcibly expressed; and this, too, strangely enough, without any distinction as to relative placing of metal; whether equally or unequally exposed, or even whether both were exposed or one completely covered.

To such extent is this peculiar discussion carried that the same gentlemen are found inveighing most decidedly against the partial filling of ordinary cavities of decay with tin, and the covering of this metal with gold, upon the ground that the *union of two metals in one cavity* would produce pulpitis, death of the pulp, and eventually alveolar abscess, and, at other times, are equally strong in their advocacy of the employment of sheet-lead as a pulp-protector (!) in cavities containing almost exposed pulps, and the filling over *it* with gold. This appears to have been done with the idea that lead was not a metal!

Views upon this "two-metal question" had become modified by time and observation, until it was some time ago announced —" Cosmos," Nov., 1879, p. 626 — that observation had shown "that amalgam and gold *may be used* in the same tooth, *and in immediate contact*, with no unfavorable results, notwithstanding the theories which have obtained to the contrary "— the italics are mine; while to a question regarding the consequences of filling two proximate cavities respectively with gold and amalgam, contact being permitted, the same speaker replied that he would "*not have allowed them to touch.*"

The deduction is that it *might be* that no unfavorable consequences would ensue, *even though the metals were in contact;* but that contact should be avoided, if possible.

During the same discussion, reference is made to my claim that "the relation of the two is the salvation of the tooth," which I do claim most decidedly.

I would suggest especial reading of the remarks of Dr. Bonwill,— same discussion, page 631,— for it is very well known that this gentleman is one of the most thoroughly informed electricians of the dental profession.

His view is, that when gold touches amalgam in proximate teeth, the teeth should be so pressed apart and the fillings so contoured as that "*when they come together again no ordinary attrition will cause them to separate.*" With this I agree.

It is twenty-seven years since my attention was directed to the systematic investigation of this subject. I was led to this by the placing under my charge of a tooth — right lower wisdom — which had a large cavity upon the articulating face, that had been filled for nearly twenty years with gold, and another upon the buccal face, which had been filled three times during the same period with the same material. The patient was then fifty-nine years of age. I suggested the trial of amalgam in the cavity from which the gold had so frequently failed, and the suggestion was accepted.

A few days after the operation, the patient returned with the statement that a metallic taste was frequently perceptible, and that occasionally during mastication a painful shock had been felt in the filled tooth.

Upon examination and experimentation, I found that during laughing the cheek so pressed upon the two fillings as to make and retain gentle connection between them, giving rise to metallic taste; and that during the mastication of a piece of cracker an occasional sudden and forcible connection would be made which caused a shock.

It occurred to me *to connect* the fillings as the first means of preventing shock; and I informed the patient that if metallic taste supervened, it would pass away with the discoloration of the amalgam filling. This connection was made by cutting through the small intervening portion of enamel,— both the cavities were large,— and the sequence proved the correctness of the prognosis; there were no more shocks, and the metallic taste became less each day until it ceased altogether.

But another result in due time attracted my notice. The gold filling upon the articulating face was slightly defective, sufficiently so for me to request an occasional opportunity for its examination. Year after year passed, and seventeen years after — patient aged seventy-six — I removed this tooth with my fingers, it having loosened from combined excessive use and gum and process recession, with the amalgam filling credited with its seventeen years of service, *and with the gold filling no more defective than it was seventeen years before.*

It was as the sequence of watching this filling and others resultant from this experiment for five years, that twelve years previous to this extraction I had ventured upon the statement to the class of the Philadelphia Dental College that I thought it probable that "the relations of metals in bi-metallic fillings gave to such fillings *therapeutic* value, and *that this value was dependent upon the contact of the metals and the exposure of both metals to the fluids of the mouth.*

It was this belief which led me to "guard" with crescents of tin the cervical walls of cavities which I proposed filling with gold. It was this which led me to fill the crevices around failing gold fillings, especially at those "vulnerable spots" which were inaccessible, and under gums, with amalgam. It was this which made me feel warranted in saying at the "New Departure," meeting, of the New York Odontological, Nov. 1877, that I had, "up to April, 1877, 1053 gold fillings of my own, and many other operators', doing service *only* because they had been guarded in this way" by tin, gutta-percha, or amalgam.

Believing that the injury to soft dentine which would arise from contact between it and gold by difference of potential would be notably counterbalanced by contact again between gold and amalgam, I can readily see that, as has been stated by Dr. Dixon, amalgam fillings might be serviceably guarded by gold; while, upon the same principle, gold fillings would be guarded by amalgam, as instanced by Dr. Daniel Neall; but upon principle I should expect in a very soft tooth, one that markedly *needed help*, that a gold filling circumscribed by amalgam would do longer service than an amalgam filling circumscribed by gold.

But it is not alone for the simple maintenance of integrity in any one or more fillings, that I would advocate the making of bi-metallic fillings or the modifying of fillings already made of one material so as to make of them "bi-metallic;" for my experience in great numbers of mouths points to the conclusion that such fillings may be definitely relied upon for the making of such change in the oral fluids as will be protective of other fillings, preventive in some degree of caries, curative of sensitiveness of dentine, and even beneficial to general health.

It is with much deliberation that I have come to these conclusions; and yet it is not without some hesitancy that I have spoken of them from the lecture-stand, and presented them for professional consideration. The instances in which the making of six or eight such fillings in mouths where failure of fillings was a constantly recurring thing, and where exquisite sensitiveness of dentine was the unfortunate concomitant, and where such failures became from that time notably less, and markedly in connection with the gold fillings, and where the sensitiveness of dentine was modified even to almost complete normality, are so numerous in my practice that I could not overlook them even if I would. Indeed, these results have been so notable, that in many mouths containing partial dentures upon gold plates, I have supplemented my efforts at preserving the remaining teeth with plastic fillings *by breaking a tooth or block from the artificial work, and building in its place a tooth or block of amalgam.*

The change, both local and systemic, which so soon occurs, is too frequent and too decided for me to doubt that the *post hoc* is a *propter hoc*.

Fractures of tooth from fillings, either of gold or amalgam.—With the classes of teeth which require large fillings, it is no unusual thing to find cavity walls of paper-like thinness, or portions of tooth possessing considerable thickness, but attached to the roots by narrow connections. In such cases a violent contact with unyielding material, such as crackers, candies of certain kinds, pieces of bone, or oyster-shells, etc., will inevitably crush or break off such fragile or insecure structure.

These accidents are probably most frequent with bicuspids, especially those of the upper jaw. This is due to the marked predisposition of these teeth to decay largely, and in such manner, mesially, distally, and in the articulating face, as to afford, after filling, the least remaining strength for resisting the disadvantageous wedging of food between the buccal and palatal cusps.

It is to avoid this contingency that Prof. D. D. Smith has suggested the advisability of excising the palatal cusps, and, in very rare cases, even both cusps, of such frail teeth, and either restoring contour with filling material, or so filling the teeth as to make of them thickened cuspids. This operation will be recognized as founded upon correct principle, and is rendered peculiarly feasible by the use of plastics, in which, although "combination" fillings are almost universally indicated, amalgam nevertheless performs the most important part.

It will be noted that such teeth are completely "slotted" through; the remaining walls are necessarily thin, and it is important that a good color should be maintained, at least, buccally. For this purpose the buccal wall is lined with oxychloride of zinc, and the buccal cusp is so undercut as to afford the strongest possible hold for the strongest kind of amalgam,—contour,—with which the tooth is filled after having secured the cervical portion of the cavity by a thin layer of "submarine." If it has been deemed desirable to cut off the buccal cusp, in consequence of exceeding frailty at the bucco-cervical line, it will be necessary to finish by the use of front-tooth amalgam.

But it is in the repairing of *ordinary* fractures of tooth substance from fillings of gold or amalgam, that amalgam most positively demonstrates its extraordinary capabilities; for, in such cases, *from the least to the greatest*, the reparations are made *with almost equal facility*.

Every indication is met, no matter how diversified, no matter how numerous, no matter how imperative! If a mere "shell" has been "artistically" filled with gold, and a long sliver of enamel has been broken out from the disto-palatal wall, a more difficult or annoying piece of repair, with gold, could not well be imagined, and yet, with soft-mixed contour amalgam, a repair can be made by "cold-soldering"—see Technicals—which will be so serviceable as to probably be in good order as long as any other portion of the filling, and it can be done with perfect facility, with comfort to the patient, and in ten or fifteen minutes.

If a large, strong molar has been handsomely and permanently—accidents aside—filled with gold, and, as the result of a tremendous bite, the whole buccal face of the tooth has been fractured off, a repair of gold would be far more tedious, far more difficult, and *far more expensive* than was the original filling. With contour amalgam, this repair would be, comparatively, but a trifle. It would take no more time than would an ordinary filling; it would be done without physical tax to either operator or patient, and would be *far less expensive* than was the filling to which it would be built on.

If small pieces of tooth are fractured, an opportunity for making, of the remaining filling, a bi-metallic, is offered. This is done so nicely, with such celerity, and with so little cost, as to most favorably impress patients with the working of *plastic dentistry*, and thus, by "little things," prepare them for continuances of relief from infliction, until they are shown that the resources are almost endless, and until they know, "of their own knowing," that good, substantial, compensating dentistry is *not* necessarily attended with the non-compensating inflictions to which they had been, for years, subjected.

Split Teeth.—For more than forty years attempts have been successfully made at rendering teeth, which had been split through the roots, useful by "bolting." The making of the gold bolt with its screw and burr; the countersinking and covering of the screw-head by filling; the after preparation and the final filling of the split root and tooth with gold, constituted a series of manipulations which made the operation difficult, delicate, and very expensive. All this is changed, and, although the work as now done with amalgam could be done with gold, and beautifully done, and could not be done nearly so well with the amalgams almost universally employed, it can yet be *better* done, taking everything into consideration, by the "amalgam method" of the present day than by any other way.

If the split tooth contains a vital pulp, the devitalization of this is the first thing to be done. The parts are gently drawn into apposition by means of book-binder's thread passed around the neck of the tooth, and an application is carefully punctured into the pulp. This application should be very small in quantity and should be largely acetate of morphia paste.

If possible, it should be *only* this, and especially for the bulbous portion of the pulp, and *if* the arsenical application is found necessary, it should be made as intimately mixed with cotton wool as possible, instead of being introduced loosely in paste form. The "devitalizing fibre" is admirably adapted for this purpose, to be used in very minute pieces. It must be remembered that the split is a direct route to the peridentium, and that the least arsenical impress upon that membrane at so high a point will probably eventuate in death of the tooth, and will be naturally followed by exfoliation.

The pulp having been devitalized, the tooth is placed in the same category with all split pulpless teeth, for which the first indication is the accurate joining together of the parts. This is accomplished by ligating at the neck with book-binder's thread, drawing it as firmly as is consistent with comfort, and having the knot held by pressure from a suitably shaped serrated plugger, until it is securely tied. The thread should then be allowed to shrink from moisture, and in ten or fifteen minutes a second ligature should be tied as firmly as possible either above or below the first ligature. The position of the second ligature is decided by the manner in which the parts draw together. The second ligature should then be allowed to shrink, when a third ligature should *replace* the first one, or be passed around the tooth on the opposite side of it from the second one.

The objects which must be gained by the ligation are, first, the accurate placing of the parts in apposition, and, second, the *firm* maintenance of this condition. The tooth is then carefully entered by drilling; the pulp cavity is cleansed, and the canals—enlarged by Talbot reamers—are, if possible, utilized for effecting the reparation. In first bicuspids and upper and lower molars, these are generally available, but in single rooted teeth the uniting drillings have to be made according to circumstances. The principle of securing is by "dumb-bell" or "double-headed" union; a drill-hole is made in each portion of split root, or the canals are enlarged in bi-rooted or multi-rooted teeth; these large, round drill-holes are then united by narrow fissure-drilled canals. The cracks of fracture are opened from the pulp-cavity almost to the periphery, and then the drill-holes, fissured canals, and cracks of fracture are all filled with contour amalgam. It is sometimes possible to add dove-tails cut in the crown portions of the teeth, but, if this is not practicable, drill-pits should be made in the substance of each part, and these should be enlarged into spheroids, that thus increased strength of union be given to this part of the filling. If there is not substance sufficient for this in consequence of previous extended decay, drill-holes should be made *completely through* the enamel; these should be countersunk upon the enamel surface, and thus "rivet-head" fillings be made.

From what has already been said, it will be at once suggested that crown-linings and rivet-heads exposed to view are made of front-tooth amalgam, while all the other filling should be of contour amalgam.

I have quite a number of these operations in *hard* service, many of which have been doing duty for from four to nearly ten years, and, as a rule, they have thus far been more satisfactory than the "bolted" operations, though some of these have done remarkably well; but the decided advantages of the amalgam "joining" are the greater facility with which the work is done; the almost complete avoidance of irritation during dental manipulation, and the reduction of expense to one-half, and in large cases to one-quarter that of the "gold bolt" operation; while the compensation for time of operator is equally great and the comfort and satisfaction to the patient are increased.

The ligatures should remain on the tooth for a day or two, if there is no gum irritation from them; but if that one nearest the gum irritates, it should be removed upon completion of operation. In a couple of days the ligatures should be cut off, and any finishing which may be required in parts covered by them, should be looked for and done.

Perforated Teeth.—These are teeth in which decay has so progressed, or in which drilling or excavating has been carried so far, as to have made an opening through the cementum; they are frequently regarded as not amenable to treatment, or, if attempted, are usually considered as extraordinary in the line of dental manipulation. A trial of some such cases, with plastic fillings, will soon make of them, to all practitioners, that which they are to the plastic workers, viz., matters of very little moment. Cotton wool, saturated with oil of cloves, is pressed gently into the cavity, and allowed to impinge upon outside tissue just sufficiently to permit an accurate contour healing—neither concave nor convex. The cavity being then *very gently* dried with *absorbent cotton*, a smooth piece of *low-heat* gutta-percha—white—is laid over the orifice and secured in position by accurate pressure—oiled instruments, cold—against the edges. This is then firmly fixed by submarine amalgam—soft mix—with which all the floor of the cavity is covered. This removes all "perforation" complication, and reduces the work to ordinary, with the exception of regard for possible future contingencies. If the perforation is very large, the tooth in the lower jaw, and the patient of strumous or otherwise doubtful diathesis, all arrangements for occasional vent, if required, should be made, though my experience is, that the "extraordinary" concomitants of these cases are the ease with which they are generally treated, and the equanimity with which the surrounding tissues accept the conditions.

Loose and divided Roots.— We have now arrived at what may be regarded as the great dividing line between gold and plastic dentistry.

Up to this point, the gold-worker is able to do, *after a fashion,* all that can be so effectually accomplished, so much more easily, with plastics; but now the limit has been reached. We here enter upon a broad field of *impossibilities* to the gold-worker, and yet a field in which the marvels of plastics seem fairly to revel.

A patient applies, who, having lost nearly all the teeth, has worn a partial denture clasped to bicuspids. One of these has gradually loosened, and has finally dropped out. The roots of an adjoining molar, quite loose and decidedly doubtful, the palatal separated from the buccal, are each drilled out, and "ring-bolts" of platinum wire — made by rivet-heading one end, and forming, by round-jawed pliers, a ring upon the other end — are filled into them. These fillings having set, the roots are drawn together by a ligature passed through the rings. When they are fairly in apposition, a foundation of submarine amalgam is made, resting upon the roots, and held between the bolts. This is allowed to set thoroughly, when contour amalgam is built up *domelike* to the ligature, entering into the rings and securely binding the whole together. After this has hardened, the ligature is cut off, a clasp is approximately adjusted and soldered on to the work; a bicuspid is also added, and the piece is placed in position; then, with contour amalgam, a nice crown is readily built on, fitting the clasp to perfection, from the cervical edge of the previous dome-like building down to the now entirely covered tops of the ring-bolts.

This soon sets, the work is carefully removed, any defect of filling is made perfect by additions of amalgam, and the crown is smoothly finished off. In an hour this can be used with safety, and, as the result of such work, the patient is made "more comfortable than ever before." *So he said.*

A patient applies who has but two teeth left on either side for mastication; one of these, a lower molar, has recently been broken, and, in the fracture, the distal half of the crown, with a large and very expensive filling, has been broken off, and the distal root split apart from the remaining crown and root. In a few days the separation between the roots became enlarged from the forcing of food during attempts at mastication upon the half-crown, and such soreness supervened as to preclude further effort upon that side. The other side has given such decided evidence of trouble, from its *constant use for a week*, that relief is sought. A ring-bolt is filled into the distal root; the canals of the mesial root are widely opened for retaining hold; and, a large drill-hole having been made through the mesial face of the half-crown, it is well countersunk from the outside. The distal root is drawn back into position by a ligature passed through the drill-hole and around the bolt. A foundation is made of submarine which connects the bolted root with the mesial half, held by the canal hold; when this is hardened, the ligature is taken off and a contour crown is built on, around, and through the ring-bolt and through the mesial drill-hole, and is rivet-headed in the countersink.

This has been eaten on, most satisfactorily, for several years.

A patient applies, seventy years of age, who has lost nearly all teeth except the superior and inferior incisors and cuspids, and these are worn almost to the gum. The left lower second molar has advanced quite to the position formerly occupied by the first molar. A crown has just been lost from the upper jaw, which, *though rather loose*, has subserved the purposes of mastication "much better than nothing." It is shown, and is found to be the crown of the upper second molar which had decayed from the palatal and mesio-buccal roots, which had gradually loosened and come away. Another root, the disto-buccal of the first molar, is in position. These three distinct and separated roots of *two teeth* are drawn together and utilized by crowning, and the patient is "glad the old tooth came out;" for he eats "so much better with the new one."

A patient applies whose eating has been done, for a number of years, by two molars of the right side and by two first bicuspids of the left side. There are no other teeth or roots on the right side; but there is the second superior bicuspid of the left side, the root of the second inferior bicuspid — loose, — and a lower molar root, quite near to it and quite firmly implanted.

The crown has recently been broken from the first inferior bicuspid, and, the eating having necessarily been done exclusively upon the right side, the two molars have begun to complain. The three roots are ring-bolted, and a long, *continuous* crown is built *upon the three*. The two firm roots hold the middle loose one strongly in place, and *both the bicuspids* of the left superior jaw *are utilized*. Ample masticating surface is thus given, and the patient relieved from the burden of artificial work.

It is needless to multiply cases, for these will suffice to demonstrate *how it is* that the plastic-worker can take the wrecks alike from gold-work and from neglect, and make the patients "rejoice and be glad" at a time of life when they have come to realize the impossibility of enduring further infliction, the folly of continuing needless neglect, and the almost boundless signification of the phrase, "*comfort from plastics.*"

ARTICLE XIII.

GUTTA-PERCHA.

THIS material was offered for dental consideration about forty years ago (about 1850).

It was suggested as a *temporary stopping* for frail teeth, and was recommended for its ease of manipulation; its non-irritating and non-conducting qualities; its insolubility in the fluids of the mouth, and its reasonable resistance to attrition.

These desirable characteristics caused it to find favor in a limited degree, even though its color — a dark brown — was very objectionable; but, upon the introduction of a mixture known as "Hill's Stopping," and supposed to be composed of gutta-percha, quicklime, and pulverized silex (a supposition probably entirely erroneous), it was quite promptly accepted with marked favor, and soon took position as a very valuable adjunct in practice.

Its inventor, Dr. Hill, had quite exalted ideas of the value of his invention, for he stated that while he did "not expect it to supersede gold entirely," he nevertheless believed that it could be "advantageously substituted for that material in many instances."

The advocates of the "compatibility theory" believe, at the present day, that his views were correct, and that if they had been very much more universally accepted, the result would have been a vastly better record in the saving of teeth.

Other compounds, of a like nature, were soon introduced as "superior," but with little or no just claims to such distinction, as, after trying a great variety of ingredients for the purpose of whitening the gutta-percha without seriously impairing its toughness, nearly all the better grades were, and are, composed mainly of *gutta-percha* and *oxide of zinc*.

Like other manufactured articles, it was soon found that *much more depended upon the art of manufacture than upon the components of the filling material*, and even yet the relative quality of the various "makes" is due largely to this.

Early in the use of gutta-percha stoppings it was noticed that "remarkable results" were being attained, and those who were investigating these with a view to the future modification of practice came in due time to regard them with exceeding favor.

It was gradually accepted that gutta-percha had its place not only as a "temporary filling," but as a stopping of *extraordinary permanency*, having but two demerits of note, viz.: inability to resist the attrition of mastication, and a degree of shrinkage which permitted *leakage* and a consequent "clouding" of the filled tooth.

The first of these objections was only to be overcome—*in the use of the one material*—by restricting its employment to such cavities as were so situated that but little or no attrition would result from mastication, and by directing especial avoidance of friction from the brush during the cleansing of the teeth.

The second objection was by no means so easily disposed of. In fact, after much thought and experimentation, it has been found practically insurmountable.

Solutions of gutta-percha were tried as "linings" for cavities more than twenty years ago, but the shrinkage from evaporation was proved to be even worse than that from cooling, and, as it was indispensable that the material should be heated in order to be properly introduced, it was finally announced that the best gutta-percha fillings leaked, but that this leakage was not detrimental *except so far as it permitted slight discoloration*.

The gutta-percha filling materials of the present day are divided into *three* grades:

First. Those of "low heat,"—having sufficient plasticity for manipulation at temperatures ranging from 140° to 200° Fah.

These are always to be warmed either over water or upon the upper plate of the dry heat G. P. and instrument-warmer, and are particularly applicable for covering almost-exposed pulps to prevent danger from thermal irritation, if metallic filling material is utilized, and as a *first layer* of filling in "deep-seated" cavities of decay. They should be worked with serrated instruments slightly warmed or cold and oiled.

They should not be used for "outside work," as their resistance to attrition is quite moderate.

I refer here *exclusively* to "low grade" *white* gutta-perchas, for it is never the case that *red* gutta-percha base-plate (a very useful and *eminently serviceable* filling material) is of higher grade than 150° or 180° Fah.

Second. Those of "medium" grade,—such as become plastic at temperatures ranging from 200° to 210° Fah.

At the present time this grade of gutta-percha stopping is confined to the "shaded" material—yellow and blue—and these, though they are somewhat lowered in "heat grade" by the work incidental to the thorough incorporation of the coloring ingredients, nevertheless possess sufficient consistence, toughness, and resisting capacity to subserve the purposes of an excellent and durable filling.

Third. Those of "high heat,"—which do not become sufficiently plastic for manipulation at less than from 216° to 230° Fah.

As these will not soften over boiling water, they have to be heated upon a metal or porcelain plate, or over the flame of a spirit-lamp. This requires much care, as it is essential to good results that they be *very gradually heated* lest they suddenly swell and deteriorate, and that they be not overheated, as then they disintegrate and become "heat-rotted."*

Much of the obloquy which has attached to gutta-percha work has unquestionably been due to ignorance of these facts, and to consequent mismanagement of the material; and I have never yet conversed with any gentleman that denied the *permanency* of gutta-percha fillings who did not overheat his material either upon a plate of porcelain or metal, or, worse yet, over the flame of the spirit-lamp.

I would state here that the value of gutta-percha stopping is not to be determined alone by the "heat-test," for *it is easy to raise the material to any reasonable degree by simply increasing the relative quantity of inorganic admixture,* but this very increase is destructive to value by overloading the gutta-percha.

---

* Only small portions of gutta-percha should be placed at a time on the 'warmer," as *frequent* heating is seriously detrimental to the best stopping.

All other things being equal, that gutta-percha stopping which gives the highest heat-test with the least admixture of foreign material is the best.

Although gutta-percha has been proved to make a reasonably good stopping in cavities from which it was found impossible to exclude the moisture, it is nevertheless essential to *as perfect a filling as can be introduced* that *dryness be maintained* during its introduction.

If this is not possible, I should recommend the employment of the *red base-plate* in preference to any preparation of white stopping, warming it over water, and using *cold* instruments for its introduction, which should be *touched to an oil-pad* to prevent adhesion of instrument and consequent "drawing" of filling material.

NOTE.—The *oil-pad* is conveniently made by cutting a groove with a small corundum wheel, around the end of any ordinary *flat top* glass stopper, and then, stretching a piece of chamois-skin over the end, securing it by a ligature of bookbinder's thread in the groove. The chamois should be of two thicknesses.

After properly trimming the chamois-skin, a drop or two of sewing-machine oil saturates it, and it is ready for use.

This little device is *essential* not only in gutta-percha work, but in connection with the manipulation of the zinc-phosphates also.

In all cases where dryness can be maintained, I should prefer to use the medium and white preparations. These should invariably be introduced by means of *warm* instruments, as the necessary degree of heat for *deliberate* and *accurate* manipulation can only be retained to the gutta-percha in this manner. For this purpose a combined gutta-percha and instrument-warmer, of the dry-heat pattern, should be used.*

The "low-heat" gutta-perchas should be placed upon the upper plate; the "medium" and "high-heat" gutta-perchas upon the middle plate, and the instruments upon the lower plate—this insures that they shall be *hotter* than the gutta-percha, and thus that they shall be as warm when taken from the plate *and carried to the mouth* as *was* the gutta-percha when taken from its place upon the other plates.

Another advantage, and a very great one, which is attained by an instrument-warmer, is the ability to heat at one time all the various instruments required in any given operation; this

---

* The water-bath warmers have been discarded since 1885.

Fill lamp two-thirds full. Put "Low-Heat" gutta-percha, red base plate, or temporary stopping, on upper plate; "High Heat," or shaded gutta-percha, on middle plate; Instruments, on lower plate. The heat for instruments, the handles being comfortable to work with (regulated by raising or lowering them, or the wick), makes all heats right.

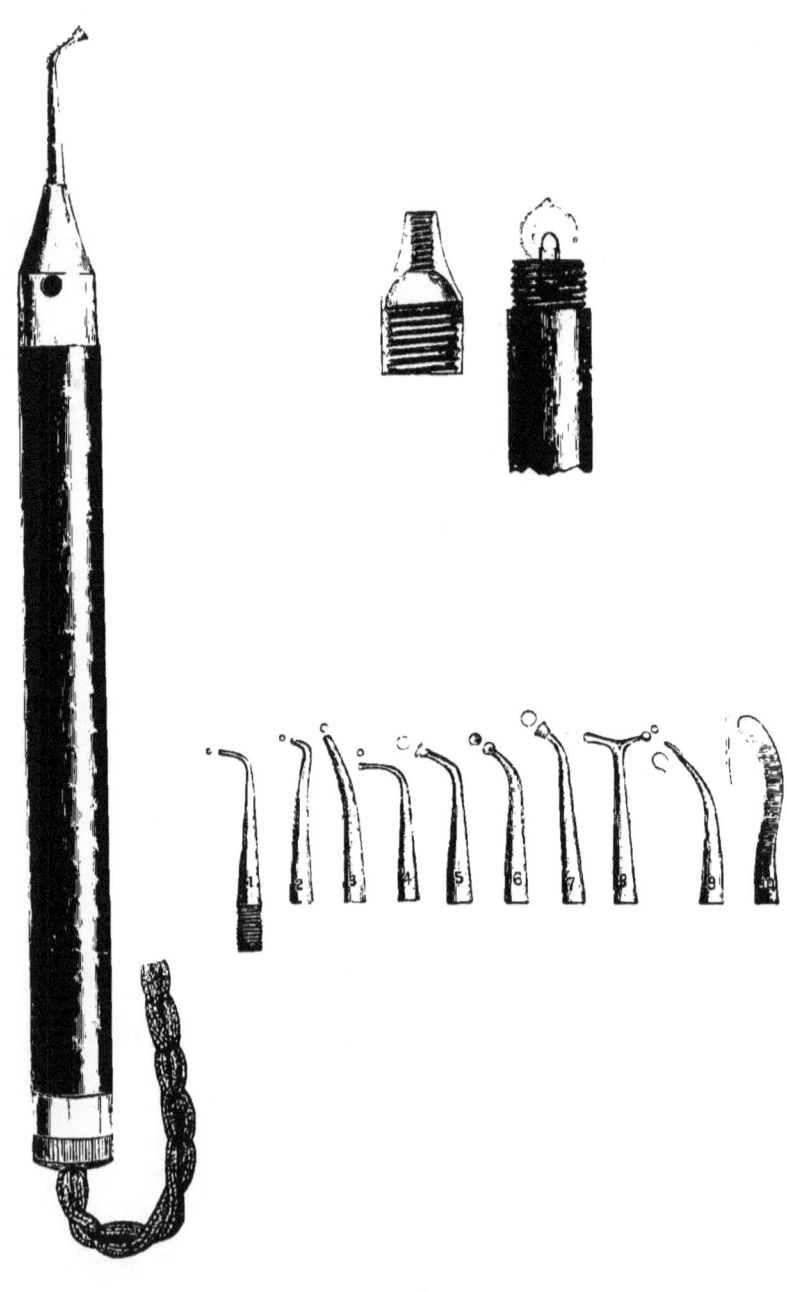

will be found quite desirable, and to result in a great saving of time.

Again, instruments heated in a flame are very liable to leave traces of soot upon the filling, while those heated upon a plate always leave it in a neat and presentable condition.*

The instruments best adapted for the introduction of gutta-percha fillings are — with the occasional exception of a ball-burnisher — *certain forms* of those used for the manipulation of cohesive foil.

It is important, however, to note that instruments intended *especially* to consolidate gold foil by the force of "direct impact" are *least of all* indicated for use in the introduction of a gutta-percha filling, for *the principles which govern the durability and value of a foil filling are precisely opposite to those which* govern the durability and value of gutta-percha fillings.

For the perfect introduction and consolidation of foil it is essential that free and fair ingress to the cavity shall be obtained, and for this purpose it is frequently necessary to cut away considerable portions of both enamel and dentine.

For the integrity of foil fillings it is regarded as better that the walls of the cavities should possess a reasonable degree of both thickness and strength.

On the contrary, for the perfect introduction and compacting of gutta-percha, it is *not essential* that nearly so free ingress to the cavity should be obtained, while *it is essential* that the largest possible portion of both enamel and dentine should be *carefully conserved.*

The *most important* of this tooth-tissue is *just that which should be cut away for foil-work*, and upon the *preservation of this* depends, almost entirely, the value of the gutta-percha work, as it guards against attrition, a material which, while it is eminently more tooth-preserving than gold, is wanting in the physical characteristic of resistance to mastication.

Again, for the integrity of gutta-percha fillings, it is not regarded as important that the walls of cavities should possess, in much degree, either thickness or strength, as it has been *thoroughly demonstrated* that some of the most signal triumphs of gutta-percha have been in teeth in which gold fillings of

---

* In *finishing* fillings, flame-heated instruments should be wiped before using.

magnificent workmanship had failed repeatedly, until the remaining walls of the now enormous cavities were almost as thin and frail as *letter-paper!*

Therefore serrated points of such *curves* and *angles* as have been found best to meet the requirements in difficult and inaccessible cavities are those recommended for the *introduction* of gutta-percha fillings.

The instruments best adapted for *finishing* gutta-percha fillings are thick and thin, convex or flat burnishers; these, together with the occasional use of a large or small ball-burnisher, will be found to meet all requirements. These should be heated in the same manner as are the pluggers.*

The filling material and instruments (both plugging and finishing) being properly warmed, the gutta-percha should be taken in small pieces — piece by piece — from its plate by means of a moderately fine probe, and thus carried to the cavity and placed in position. If practicable, each piece should be made to adhere to the wall of the cavity, and then be accurately packed into position by the appropriate plugger, until the cavity is either entirely lined, or is partly filled, when the completion of the operation is only a question of a short time; but when this adhesion of the first pieces is difficult, they should be held in position by the probe until they are made to adhere to the walls by the use of a plugger, when that adhesion should, in turn, be maintained by the plugger, while the probe is carefully withdrawn.

It is an important consideration that accuracy in amount of filling material should be regarded, as thus the minimum of surplus will remain for removal prior to finishing.

During the removal of surplus material, and the final smoothing of the filling, it should always be remembered that the work must be *towards the edges of the cavity*, as thus the filling is maintained "flush" and the gutta-percha kept close to the walls.

While finishing, it will be found that moisture is not only *not detrimental,* but in some cases rather advantageous, as it permits a smoother cutting of the gutta-percha.

In *dry* finishing, only that portion of the filling material which will adhere to the instrument *from one touch* should be

---

* Nos. 2, 3, 6, 7, and 8 of the amalgam set are heated for such use.

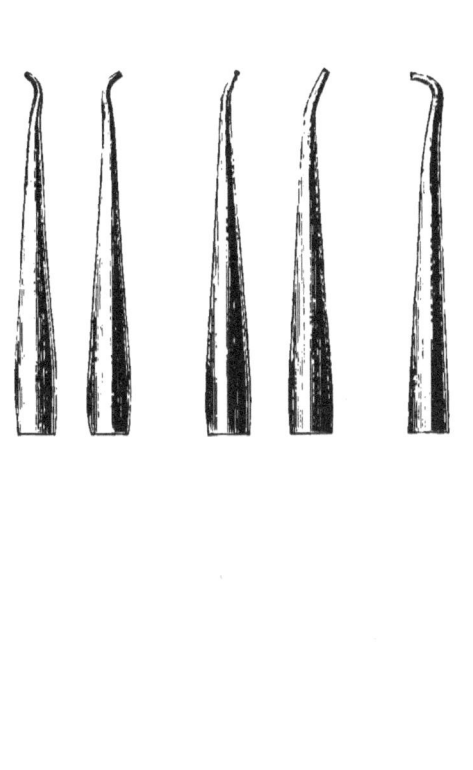

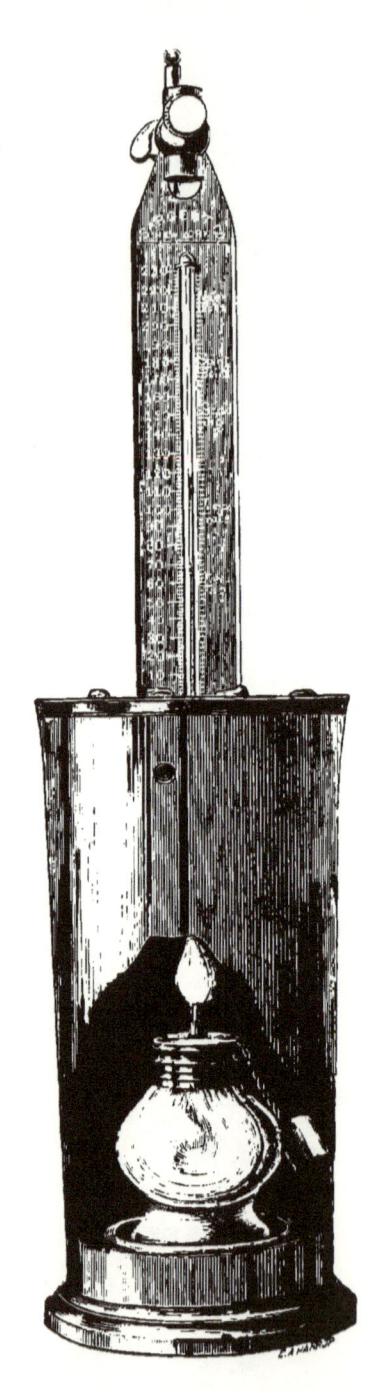

removed at a time, as retouching will complicate and retard progress rather than be productive of advance.

Tests for Gutta-Percha Stopping.—As there are very numerous "makes" of gutta-percha, and as some of these, least worthy of confidence as filling material, are advertised as "the very best" of their kind, I deem it important that reliable *tests* for this valuable aid in saving teeth shall not be wanting, that thus with care in heating, and with proper working, the excellent results which are possible with a *good material* may not only be confidently anticipated, but fully realized.

First. *Heat-Test.*—For the accurate establishment of the "grade" of any given sample of gutta-percha, it is, of course, necessary to employ a "heat-tester." The illustration sufficiently explains itself, as the instrument is, practically, a gutta-percha warmer with thermometer attached. A sufficiently accurate arrangement can be made by taking any small tin box,—such, for instance, as are bought with spices,—and cutting an opening in the cover sufficiently large to permit the introduction of a thermometer. It should then be partially filled with water and arranged for gradual heating.

The thermometer should register at least 212° F., while that used for the instrument illustrated registers 230°. No gutta-percha can be serviceably worked which grades higher than 220°, and I regard this as uselessly high. At 240°, gutta-percha "heat-rots," and does not regain its previous toughness or integrity, and if at 208° or 209° it softens sufficiently for easy manipulation, it is, if properly constituted, sufficiently resistant to respond to all the legitimate requirements of a good gutta-percha filling material.

But, as I have said, the value of gutta-percha stopping is not to be determined by the heat-test alone, for it is easy to raise it to the proper "grade-test" by addition of inorganic constituents. To further determine its value, then, it is necessary to decide upon the relative quantities of organic and inorganic constituents; for this we have the "fire-test."

Second. *Fire-Test.*—This is done by weighing a small quantity—say, a pennyweight—which it is better should be in one piece. This is then subjected to high heat; a convenient method is to place it upon a long-handled iron spoon, and put it on a hot fire; in a few minutes it takes fire; burns with a bright blaze; ceases to burn; becomes red hot and thoroughly calcined, while it perfectly retains its original shape. It should then be allowed to cool, after which the residuum should be carefully weighed.

If one pennyweight of gutta-percha stopping has been thus treated, the residuum should weigh a little over eighteen grains if the inorganics are salts of lime and zinc and the gutta-percha stopping of a "low-heat" grade—1 part G. P. to 4 parts salts. If the pennyweight tested is "high heat," and is properly and toughly plastic at 209°, the residuum—if lime and zinc—should weigh exactly twenty grains—1 part G. P. to 5 or 6 parts salts.

I refer to these proportions and ingredients because a THOROUGH KNEADING *of them upon an iron or porcelain slab at a moderate heat—*190° to 200°—*will produce an excellent filling material of the two most desirable grades.* The gutta-percha to be used is that which is sold in large, dark-brown sheets, resembling dark sole-leather, and *not* that which is light colored, and either in thick pieces or thin paper-like sheets.

For making of a small quantity of gutta-percha stopping — one to four or five ounces—it is needed that a vessel be arranged for hot water *and* heat. Upon this is placed a cover of tin, iron, or porcelain—a thick iron slab, nickel plated, is appropriate — with a tube for escaping steam. Upon the slab or cover is placed the gutta-percha and a portion of the oxide of zinc. As soon as the gutta-percha softens, the zinc powder is gradually kneaded into it. In my experience, this is best done by means of a small wedge-shaped tool of iron — very blunt — set on to a strong rod handle, turning the mass with a short, stiff spatula. The incorporation of the zinc should always be in consonance with the maintenance of the *toughness* of the gutta-percha — never faster — and, *after the zinc is all incorporated,* the mass should be *thoroughly kneaded* for an hour or more, until the stopping is perfectly *toughened.*

It is possible that better — tougher — gutta-percha stoppings may be devised in the future, but all unfounded pretensions in that direction may easily be detected by the tests given; meanwhile, those which are made from the formulæ, and by the method given, will be found reliable, as they have been "placed" by the results of thirty years of experiment, during which time the inferior stoppings that have too generally been made, sold, and used, have given so much dissatisfaction as to have caused the mass of the profession to accept as true the oft-repeated solecism, that the making of good gutta-percha stopping is one of the "lost arts!"

As in succeeding years the quality of the crude gutta-percha has become inferior to that which was first obtained, it is very gratifying to be able to say that from the provings of every test — theoretic and *actually practical* — the advance in manufacture of gutta-percha stopping has fairly exceeded in degree the deterioration of the crude material. Thus it is that to-day we have gutta-percha stopping which, with notably less proportions of inorganics than pertained to Hill's and Bevan's stoppings, "test" by heat as even *tougher* than those excellent materials, while the results of the last decade of *actual service* give satisfactory evidence of equal worth and, in many instances, promise of relative superiority.

To the discarding of the salts of magnesia, talc, cements, etc., and to the gradual introduction of the sulphite, bi-sulphite and sulphuret of lime, soda, and zinc, is unquestionably due the improvement of G. P. stopping; but with all the work that has been done, it seems evident that the *possibilities* have not been reached, so that we still say that "better, tougher gutta-percha stoppings may be devised in the future."

With every step, however, in advance, the difficulty of *determining* the value of relative proportions of the inorganics increases markedly, and thus it is evident that to those who prosecute this task it will be a life-long work.

At present it seems determined that oxide of zinc, controlled by sulphite of lime or bi-sulphite of soda, and the mixture in turn controlled by sulphuret of zinc, makes the best grade of results, and these in such proportions as will give to one part of gutta-percha four to four and a half parts of oxide of zinc, a half to three-quarters of a part of lime or soda, and from one-twentieth to one-fiftieth of a part of zinc. These, with "hand-making" instead of machine-making, and with a graded heat of about 190° to 200°, give us the material which, from all my carefully collated statistics, seems worthy of the most confidence.

Gutta-percha stopping is "shaded" by using yellow-ochre in proportions of about twenty grains to the ounce, or lamp-black, *carefully distended by admixture with the yellow powder prepared for "yellow" stopping*, in proportion of about one and a half grains to the ounce. The work which is necessary to give a uniform shading to G. P. stopping is of such continued duration as to diminish in some degree the "heat grade" of the product, and it is for this reason that the colored stopping ranks as "medium heat."

By coloring gutta-percha stopping with carmine and with jewellers' "rouge," a gum shade can be obtained with which cavities on the labial faces of denuded incisor and cuspid *roots* may be filled with decidedly artistic effect.

Durability of Gutta-percha.—It was quite a number of years after the entrance of gutta-percha into the list of filling materials before it came to be regarded by any operators as other than subservient for "temporary" work, but, after a time, cases began to multiply in which fillings inserted *avowedly* as "temporary" continued to do good service in such manner as to excite both curiosity and professional interest as to the length of their possible durability.

It was next noted, as years passed by, that the gutta-percha fillings had actually lasted, *in many instances*, longer than the gold fillings which had been previously introduced.

Results such as these, in my own practice, together with corroborative comparison of views with others, induced me to institute an extended course of experimentation in this direction.

As the result of over fifteen years of careful observation, and with a basis of over *two thousand* replacements of gold with gutta-percha, I find that, "properly used," gutta-percha is *at least twice* as durable as gold, and that in *very soft teeth*, in selected places, it is fair to presume that it will preserve its tooth *at least three times as long as well-introduced gold fillings.*

I wish it to be understood by the profession that I do not make this statement as an *opinion*, but as *an assertion* vouched for by long, carefully tabulated records, and as a basis upon which the operators of the future may found a satisfactory tooth-saving, and comfort-giving practice.

Cavities in which the use of Gutta-percha (alone) is indicated. — In former years, the cavities in which experience seemed to have conclusively proved the capability of gutta-percha for making a reliable tooth-saving filling were much more numerous than at present. At that time all circumscribed cavities upon the mesial or distal faces of *soft teeth*, as well as many cavities upon the labial faces of incisors and cuspids and buccal faces of bicuspids, and especially large cavities upon the buccal faces of molars near to and even beneath the gums, were regarded by gutta-percha workers as the places where they achieved their greatest triumphs.

Gradually, as "combination fillings"* began to be recognized as advanced practice, the legitimate use of gutta-percha alone has become restricted, first, to large cavities approaching the pulps, and having thin, frail walls located upon the buccal faces of molars and bicuspids, either superior or inferior; second, to circumscribed cavities upon the mesial or distal faces of incisors, cuspids, and bicuspids having reasonably thick walls *labially* and *buccally* (sufficiently so to prevent "clouding" from leakage), and having an unbroken articulating wall of sufficient strength to resist the action of mastication.

In *soft teeth* all these requirements are but infrequently found, and thus it is that, at the present day, the strictly legitimate use of gutta-percha *alone* as a filling material is almost entirely confined to *large cavities, having thin, frail walls, located upon the buccal faces of molars.*

But it is by no means the case that the use of gutta-percha is confined to cavities in which it *alone* is indicated as the *best* of all filling materials; for its great value as a "guard" at the cervical portion of cavities between teeth, and as an "intermediate" between almost exposed pulps and amalgam or zinc-phosphate fillings; as a complete filling for "taps" either through tooth structure or other filling material, and as filling for "lined" cavities of large size and with frail walls, so situated as not to require a resistant to attrition, has been so satisfactorily demonstrated during the past twenty-five years as to offer every guarantee that such use of it will be eminently satisfactory alike to patients and operators.

---

* "Combination" fillings are those in which two or more different filling materials are *combined* or *made to adhere* together, and sometimes to the cavity walls.

ARTICLE XIV.

OXY-CHLORIDE OF ZINC.

IT is now about thirty years since M. Sorel, having devised the combination of oxide of zinc and a solution of chloride of zinc, suggested its possible value for the "stopping of hollow teeth," as its plasticity, when first mixed, and its subsequent rapid hardening and apparent impenetrability to moisture, seemed to render it peculiarly applicable.

Within a couple of years or so after its discovery, my friend Mr. Edward Parrish — afterwards of the Philadelphia College of Pharmacy — received from Europe a sample of the material as prepared for dental use, and, calling my attention to it, gave it to me for the purpose of experimenting as to its real worth.

At first, it promised to meet certain requirements which were admitted to exist, and which it was very desirable should be better fulfilled; but it was not long before the action upon it of the fluids of the mouth was recognized as very different from that of ordinary moisture, and, although several varieties of this easily-made compound were soon offered to the profession, they were received with evident caution and with much distrust.

The time required to disprove its claims to be regarded at all favorably as a permanent filling material was not very long, as, within a year after its introduction, many failures had occurred, and in some instances the fillings had almost entirely disappeared; in fact, its proper position was accorded it with remarkable promptitude.

Recipes for its manufacture, containing oxide of zinc, silex, borax, alum, ground glass, etc., as components of the powder, together with deliquesced chloride of zinc, or metallic zinc dissolved in muriatic acid, combined with soluble glass, for the fluid, were freely given at conventions and published in the journals; but its employment has always been confined, by those best informed in regard to it, first, to the obtunding of sensitive dentine in superficial or ordinary cavities of decay, using it for this purpose as a temporary filling; second, to the occasional filling of frail incisors, cuspids, and bicuspids, for as long a period as it would last, but avowedly with the intention of frequent renewal; and, third, as a lining for cavities having thin walls, and as a filling for the larger portion of very large cavities, when it was proposed to fill externally either with amalgam or gold.

Both recipes and uses have remained *always* practically the same; the adjuncts of burnt alum, ground and soluble glass, etc., have quietly become, to a great extent, ignored; but the oxide of zinc, silex and borax, and the solutions of chloride or muriate of zinc, are the same to-day as when they were given us thirty odd years ago.

With all its deficiencies, *oxy-chloride*, under a multitude of names, such as os artificial, osteo-dentine, crystalline, rock, agate, acme, and various other cements, together with a numerous list called after their makers Houghton's, Robert's, Smith's, Fletcher's, Poulson's, Franzelius', etc., has steadily been given in response to increasing demand, until the quantity which is now used, though less than formerly, is yet considerable.

In this connection it is truly remarkable that this compound, so worthless for general purposes, so utterly unreliable as a filling material, so almost completely ignored by the "plastic-filler," should be used in such unnecessary amounts. It seems as though, in desperation, those who have patronizingly styled gutta-percha "an excellent material for *temporary* fillings," and have decried amalgam as an "abomination, used only by incapables, and endured only by imbeciles," have been forced to content themselves with this miserable apology for a plastic filling. They have *discussed it periodically*, even within the past year, repeating again and again the few things that could be said of it; they have spoken of its "proper manipulation" just as though it were not far easier to work than either gutta-percha or amalgam. It has been not only admitted to be a good thing under gold, but a most excellent device, "well worthy of much consideration;" this, however, was not so regarded until after the plastic-fillers had been so using it under gutta-percha and amalgam for *more than a dozen years!* It was finally "accepted" *concomitant with a modified oxy-chloride* — so stated — but the plastic-fillers *never knew* of any such modification; and thus, as the plastic-fillers had gradually, *most completely ignored* oxy-chloride of zinc *as a filling material*, the gold-workers had come to using it by the *hundred-weight*.

Various efforts have been made for the "improvement" of oxy-chloride; ground feldspar, pulverized French chalk, oxide of tin, and many other like ingredients, have each been tried and abandoned. The most of these were found to act prejudicially, and none of them produced any decided improvement over a well-prepared, properly calcined "Sorel cement."

Some nine or ten years ago a method was proposed for treating oxy-chloride of zinc fillings by rubbing them with heated talc; points of French chalk (talc) were secured in crayon-holders, and, after being heated over a spirit-lamp, were used to smooth, dry, and harden the surface of the fillings. These were afterwards burnished with agate burnishers, and a peculiarly beautiful effect was thus produced.

It was found, upon very limited trial, that the whole thing was deceptive, and that there was no value in it. The first fillings which I introduced and finished by this process — 1876 — began to show evident signs of wear in less than a month, and in less than a year positive proof of its general inutility was afforded to such extent as to warrant its total abandonment.

It is unquestionably true that some of the oxy-chlorides are better than others, for they are more thoroughly compounded and more carefully prepared; but it nevertheless remains that the best of them is entirely unreliable, and that no one of them is deserving of rank as *a filling material*.

But oxy-chloride has great value for the plastic-filler as a "liner" of cavities prior to filling. It was early noted that the doubts which were expressed as to the possibly deleterious effects of the material upon tooth-bone were proving more and more groundless. It was early noted that the "chalky" condition of dentine, which was confidently prognosed as likely to ensue, did not appear, but that, on the contrary, *this structure was maintained in singular integrity*. As year after year passed by, it was noted, with increasing interest, that *decay was never seen under oxy-chloride of zinc*, and thus it was that in "plastic dentistry" it advanced from the position of a mere foundation—an advantageous, non-conducting, occupant of space—to the dignity of a complete "liner," a preventive alike of discoloration and decay.

It is only ten years since Prof. D. D. Smith read the paper which *first* openly advocated the use of oxy-chloride "linings" from the *therapeutic* stand-point; a matter of vital importance, a point for grave consideration, and one which it seems to me cannot be controverted.*

It is true that oxy-chloride of zinc remains, most persistently in some cases, doing duty as a filling, where everything else seems to fail; but these instances are so very rare, and the indications which point to this possibility are so entirely unknown, that this fact alone stamps the filling as the *most unreliable* of all, as it will *probably* fail very soon, but *may possibly* last for many years.

Oxy-chloride of zinc fails in two ways; *first*, and least notably, from attrition; for this reason it usually lasts best in cavities upon articulating faces, where it sustains the wear incident to mastication; and, *second*, and most notably, from solution or disintegration at the cervical portion of the filling. At this point, *it* fails as decidedly as does tooth-bone — soft structure — at the cervical margins of metallic fillings, especially those made of gold.

This matter of *respective* failure, in this connection, is most interesting and instructive to the student of the "compatibility" theory; for, observation shows, that with gold as a filling material, the cervical margin *of the filling* remains intact, while the contiguous tooth-structure is markedly decayed; and that with oxy-chloride as a filling material the cervical portion *of the tooth* remains comparatively intact, while *the filling* is rapidly disintegrated.

To the plastic-filler, whose thoughts are ever upon tooth-salvation, this fact alone would decide the relative value of the two materials as tooth-savers, and he would rather renew fillings as often as required than permit successive decay to finally destroy the tooth; for, he reasons, filling material is obtainable in any quantity, but tooth-bone is limited in amount, and teeth are limited in number; and yet he does not stop at this decision, for experience has taught him that other materials are

---

* Read August, 1878; published in "Cosmos," October, 1878.

infinitely more reliable for filling at this point — materials like submarine amalgam, tin, or, perhaps best of all, gutta-percha. And therefore it is, that although in some mouths, where oxychloride has been proven to do well, a plastic-filler *might* use it for an articulating filling, yet he would never think of employing it for filling approximal cavities or cavities approaching near to, much less impinging upon, the gum.

From an experience of thirty years, I am taught *never to use oxy-chloride of zinc as a filling material;* that is, as a material for filling, entirely, a cavity of decay; or for filling a gutta-percha or varnish-lined cavity in such manner as would leave the oxy-chloride exposed to the action of the fluids of the mouth.

But, as I have stated, this material is regarded as invaluable for use as a "lining" to cavities having thin, frail walls, and it is for this purpose, *and for this only*, that I ever use it.

Oxy-chloride of Zinc, as prepared for dental use, consists of a *powder* and a *fluid*. Portions of each, being mixed together into a thick creamy paste, soon form a reasonably hard, whitish cement. The powder is made by thoroughly rubbing together, in a small mortar, 3 grains of borax, 6 of silex and 1 ounce of oxide of zinc. I give these proportions in troy weight, because the weights sold with dentists' scales are usually of that denomination. These are then placed in a small crucible, and subjected to a *glowing red* heat for half an hour; by this means we have both powder and "frit," which, when cool, is finely pulverized by again rubbing it in the mortar.

This preparation of Oxy-Chloride of Zinc is adapted expressly for "lining" cavities of decay, prior to filling them with gold, amalgam, or gutta-percha. It is, therefore, carefully calcined, so as to make a "slow-setter," but will eventually become *very* hard, if mixed as directed. The object of the slow setting is to permit *deliberate* and *accurate* "pelleting," that thus perfect lining may be attained. The needs for extreme hardness are *strength* and *durability*. The zinc is calcined to free it from the moisture which it absorbs when exposed to the air, and the powder should be kept in a tightly-stoppered bottle; this can be afterwards shaded, if desired, by the addition of yellow ochre.

The fluid is made either by saturating muriatic acid with metallic zinc — allowing the solution to clear and then decanting — or, by dissolving half an ounce of chloride of zinc in two and a half to *three* drams of water; the latter is the easier and, I think, the better method. Considerable heat is generated by this dissolving, therefore the bottle in which it is done should not be too tightly stopped. After several shakings, the contents should be allowed to settle for some days, when the perfectly clear portion of fluid should be poured off for use. If it does not clear, filter the solution.

For the mixing of oxy-chloride of zinc, and, indeed, any of the zinc plastics, a piece of plate-glass, one-fourth of an inch thick, two or two and a half inches wide, and three or three and a half inches long, will be found very convenient and desirable. Upon this is placed one or more small portions of fluid, each portion separate from the others, distant an inch or so; a sufficient quantity of the powder is then poured upon the slab, and the materials are ready for mixing.

It is usually directed that the cavity be *filled* with oxy-chloride, and that a sufficient length of time be given for the complete hardening of the material — a day or more — when, it is said, it should be burred out as desired, and the covering filling of gold or amalgam be made. These directions, together with the teaching that the *main* objects accomplished are the securing of a non-conducting medium which largely fills the cavity, and thus prevents the necessity for so much irritation to the surroundings and so great expenditure of time in the introduction and consolidation of the gold filling, are not at all in consonance with the ideas of a plastic-filler, for it is recognized that oxy-chloride of zinc is a *notable shrinker;* that it begins to shrink in a short time after it is introduced, and that it continues shrinking for several days; that this shrinkage must result in a defective lining, and it is therefore taught, from the "plastic" stand-point, that the oxy-chloride should be introduced in small portions, and that it should be pelleted — see "pelleting," in Technicalities — into position against the walls of the cavity.

Furthermore, it is not regarded that the *main* objects secured are non-conduction and the filling of the cavity, in large degree, with a material which will prevent undue irritation from dental manipulation, for in plastic filling oxy-chloride is *never* used for its non-conducting characteristic, and is *never needed in bulk*, but is rather desired in its thinnest possibility, as obviating shrinkage, and none of the materials used in filling, necessitate any dental irritation.

In plastic dentistry, the *main* objects which are regarded as secured by the "lining" with oxy-chloride are, first, *exemption from further decay;* second, maintenance of color, improvement of color in discolored teeth, and prevention of tooth discoloration from filling materials liable to produce such result; third, the securing of *solid support* for amalgam, when gutta-percha — a yielding material — is used as a non-conducting pulp protector or guard for cervical edge of cavity; and, fourth — incidentally — as a strengthener to frail walls which would be liable to fracture during the progress of "treating" a tooth.

Oxy-chloride of zinc, though used with impunity in pulpless teeth,— pulp cavities filled with gutta-percha,— should be used with much caution in teeth of soft structure and non-recuperative attributes, containing living pulps. Its introduction into cavities in such teeth should *always* be preceded by the careful placing of temporary stopping or rubber varnish—virgin rubber, grs. v; gum mastich, ℈ij; chloroform, f℥iij—on muslin, as, by these means, pulp irritation, from the chloride of zinc fluid, is prudently guarded against.

Directions.—To sufficient liquid add about an equal bulk of powder. Spatulate to a soft cream, then add powder, little by little, *spatulating to thorough softness with each addition;* continue adding powder until a decided putty-like consistence is obtained; place in position by point of spatula, and "pellet" with *previously* prepared pellets of cotton; dry with bibulous paper, if needed.

It will set in from fifteen to twenty minutes sufficiently to permit of the introduction of any plastic filling, but should be covered with "Temporary Stopping" and allowed to harden for twenty-four hours if a gold filling is to be introduced.

The experience which I have had in the direction of "lining" teeth with oxy-chloride, makes it seem to me strange to hear the *therapeutic* action of filling material denied, and still more strange to see the constant presentation of certain qualities as "resistance to the fluids of the mouth," "resistance to attrition," etc., as characteristics of *a material* which are *essential* to the *saving of teeth*. *The idea is ever tacitly accepted, that a cavity of decay must be filled with* ONE MATERIAL *which shall combine all the varied attributes needed for success, totally ignoring the palpably presented fact, that materials which possess certain tooth saving attributes, and are deficient in others, can be utilized* IN THEIR PROPER PLACES, *while these, again, can be protected by other materials, which, though deficient in essentials possessed by the former, are, in wonderful degree, possessed of the very essentials which, in these, have been found wanting.**

I have hundreds of cavities, in the teeth of my patients, from which gold fillings, introduced by some of the very best operators of our country, failed from surrounding decay, and literally *dropped out*, in from two to five years, and in which amalgam fillings had been tried, some of them by the same operators, with but little better success, in which "linings" of oxy-chloride, protected by amalgam coverings, have done good service for from six to twelve years, and are yet in such condition as bids fair to afford years of service during the future.

I cannot dismiss the subject in hand without referring specially to oxy-chloride as a restorer of color to discolored teeth. It has been my experience that discolored teeth in which "bleachers" have been used, gradually lose the renewal of beauty which had been afforded them, and that the after discoloration is often worse than the first. For this reason I have long abandoned this method of "whitening" teeth, only resorting to it for an occasional experiment, as various materials or combinations have been successively recommended. Of all these, I have never yet tried any one which has proved so satisfactory, and has maintained so generally a permanent effect, as oxy-chloride of zinc.

Quite early in its use, the capability which it possesses of producing a marked *whitening* of teeth was duly noted. In many cases of exceedingly frail front teeth, with very thin

---

* Thus making a "composite" filling, in which two or more different filling materials are used *without union* between the materials.

labial enamel, its introduction caused such "spottings" of white *through this transparent tissue* as to render it necessary, for beauty, that the filling should be removed and replaced by an oxy-chloride which had been *shaded by yellow ochre* expressly for this purpose.

It was an easy deduction that thus indicated the worth of a *white* oxy-chloride as an adjunct to the old-fashioned prepared chalk and the scraped French chalk — talc — which have been used for half a century to remove the heavy discoloration prior to "bleaching," and which, with me, have still retained their value; therefore, I would now recommend, as the result of twenty years' experience, that the *whitening* of teeth — I discard the term "bleaching" as inappropriate — should be accomplished by first entering thoroughly and removing, as much as possible, consistent with maintenance of strength of tooth, all discolored dentine, especially at the cervical portion; then, after proper treating, the canal should be filled for three-fourths its length from the apical foramen. If this is done with cotton, it should be perfectly protected by temporary stopping.

The cavity should now be packed with prepared chalk or pulverized — scraped — French chalk, and this should be covered either with small pellets of cotton, tightly packed, dry, and afterwards moistened with varnish of gum-sandarac, or with an oxy-chloride or zinc-phosphate capping. The chalks act to remove the organic discoloration just as they do in the usual household application of these materials for removal of such stains from dress goods, floors, carpets, etc.; and, indeed, I was told, more than thirty years ago, that it was the domestic use of chalks for this purpose that first suggested their like employment in dentistry.

A few days should be given for the action of the chalk, when it should be removed, and, if indicated, a second application should be made. When the color has been approximately restored, the tooth should be placed under rubber, and the cervical portion "lined" with a *thin film of very white* oxy-chloride. The object of this thin lining is, that very little, if any, pain is induced by the attenuated oxy-chloride, while the pain which is apt to result from a bulk of this material — even in pulpless teeth — is sometimes very considerable, and may be of some length of duration.

After this lining has "set," a process usually requiring fifteen or twenty minutes, I prefer to fill the cervical portion, as far down as the edge of the lining, with gutta-percha, as this material is easier of removal, if, at any future period, it is deemed desirable to relieve peridental irritation by entering the canal.

The crown portion should next be lined with oxy-chloride, so shaded as to result, if possible, in a color *lighter* than that of the adjoining teeth. The reason for this effort is that the moisture will detract somewhat from the color attained, and the *final* result will thus be exactly that which is desired.

Oxy-chloride is usually shaded with yellow ochre, and, very rarely, with lampblack.

The tooth may now be filled with an amalgam made from one part of "contour" and two parts of "facing" alloy, if a resistant filling is required; or it may be filled with gutta-percha, if the filling is not exposed to attrition, or if the tooth has been intermittingly troublesome and may possibly require an occasional relief dressing.

ARTICLE XV.

OXY-SULPHATE OF ZINC.

THIS plastic preparation is, even more than the oxy-chloride of zinc, to be regarded as *not at all useful as a filling material*, but, like the oxy-chloride, it is a most valuable adjunct in its proper place. This also consists of a powder and a fluid. The powder is composed of effloresced sulphate of zinc, one part, and calcined oxide of zinc, two or three parts.

The powder of oxy-sulphate of zinc is made by triturating in a mortar two pennyweights of effloresced or desiccated sulphate of zinc and adding five pennyweights of oxide of zinc; rub together thoroughly; calcine at a *glowing red* heat for ten minutes; when cool, grind to powder in mortar. The powder should be kept in a tightly-stoppered bottle.

The fluid is made by dissolving ten grains of chloride of zinc in one fluid drachm of water. This makes a turbid fluid which soon precipitates and clears; this should be shaken when used.

Oxy-sulphate of zinc is probably one of the best pulp protectors and pulp cappers which has ever been used; it is easy of adaptation, quite as much so as gutta-percha, gutta-percha plaster, adhesive plaster, or the oxide of zinc and oil of cloves pad; but it possesses certain peculiarities which are very valuable, and which appertain to no other capper or protector, except, perhaps, plaster of Paris, which it very much exceeds in *rapidity of setting* and in *density* after it has become hard. Together with these attributes, it is not only perfectly *non-irritating* alike to sensitive dentine and to the dental pulp, but, from its admixed sulphate of zinc, is accredited, with much show of reason from extended experimentation, with therapeutic value as an astringent, antiphlogistic pulp preserver.

This is mixed upon the glass slab already described, much in the same manner as is oxy-chloride, but it should be mixed *much thinner;* indeed, it should be but little more than milky in its consistency, and certainly not thicker than very ordinary cream.

It should then be worked with the spatula until it *begins* to give the *least* perceptible evidence of thickening, when it should be taken upon the end of the spatula and placed accurately in position in the cavity by being pushed off the spatula by a moderately fine probe.

It will adhere nicely to the dentine if it is properly manipulated, and, in its still yieldingly plastic condition, can be spread, as required, by careful working with the tiny end of the probe. It should not be worked after it ceases to flow, comparatively easily, under this instrumentation.

Another method, and a very excellent one, particularly adapted to difficult and inaccessible capping, is that of preparing a pellet of cotton, *no larger than the head of a small pin*, which should be dipped in the milky zinc-sulphate *immediately* after it is mixed. The tiny capping is then gently placed accurately in position, and will be readily secured in place by touching its edges with a small, smooth-ended instrument.

It now requires a few minutes — from five to fifteen — to set sufficiently for progressive work, and then permits the introduction of oxy-chloride side linings, or of amalgam or gutta-percha fillings.

This operation is one requiring considerable dexterity, and some practice, for its acceptable accomplishment; but with these it becomes a neat and most reliable method for protecting almost exposed and *exposed* pulps.

ARTICLE XVI.

ZINC-PHOSPHATE.

THE attempted utilization of the acids of phosphorus for the purpose of compounding filling materials is now of quite respectable antiquity when compared with the still moderate age of dentistry as a profession; for the old recipe of Ostermann, published in 1832, may almost be regarded as cotemporary with the "silver paste" of M. Taveau.

This formula, for which I am indebted to Dr. De Montreville, is as follows:

"Take 13 parts of finely powdered lime and 12 parts of anhydrous phosphoric acid. This compound is moist during the mixing, and while in that condition is to be put into the cavity of the tooth, the cavity being prepared in the usual manner. The filling must be used in one or two minutes after it is made, as it almost instantly hardens. The lime must be caustic lime, and chemically pure." A second old formula, furnished me by the same gentleman, reads thus:

"Mix equal parts of finely-powdered silicate or fluate of zinc and alumina, with a sufficient quantity of water to make a homogeneous paste. This is to be introduced into the cavity of the tooth to be filled, and the drying of the mass is to be aided by warming it with heated instruments."

From these two initial cements the phosphoric filling materials of the present day seem to have gradually developed, but it is only during the past few years that the attention of the profession has been especially directed to them. The cements of Rostaigne, Poulson, Grass and Worff, Fletcher, and others—

a series of variably good materials — were quickly followed by many, more or less worthy, imitations of the genuine article. All the former are entitled to be called "ZINC-PHOSPHATES," for they have the obdurate *nitrated oxide of zinc* as their powder, which is made into paste with variously, and in some instances accurately and scientifically, prepared fluid or crystals of such form of phosphoric acid as makes reasonably durable, and sometimes remarkably durable fillings; while many of the latter are merely OXY-PHOSPHATES OF ZINC, — for their powder is nothing but calcined oxide of zinc, — a material incapable, so far as is known, of making a resistant cement; or yet worse, a mixture of oxide of zinc and oxide of tin; while the fluid is merely glacial phosphoric acid dissolved in water and either evaporated to the "desired" (?) consistence or permitted to form a natural separation, by time, and then decanted.

The former of these methods is eminently unscientific, as experiment proves it to be impossible ever to repeat the process with a known result; and eminently unsatisfactory, as none of the results make a cement which responds to a single "test" for a good "zinc-phosphate."

By the latter method, with the nitrated oxide of zinc, an approximate to a good result is produced, but with the calcined oxide of zinc only a poor cement, at best, is made.

The amount of work which has been done in the direction of these cements is something extraordinary; and it is equally noteworthy that only those whose "contributions to knowledge," in this line of experiment, have proven that they knew nothing about it, have ventured to place themselves upon the record!

There is no other plastic filling material in regard to which I have had such hesitancy in both speaking and writing, even in the familiar way of correspondence, and no direction in which I am better satisfied to be regarded as "not posted," for I feel sure that, so far, those who know most of it, recognize that they know but little, while, unfortunately for the society records, and for the members of the profession who depend upon them for information, those who know nothing of it, have freely given *all they know.*

I have been for years in constant correspondence with those

whom I had reason to suppose most earnest and best informed in this work, and time and again have we thought that "success" was to reward our efforts; but the ruthless evidence of deteriorating fluid, and crystals, and disintegrating fillings has repeatedly dampened our ardor until it has wellnigh, at times, extinguished our zeal.

But ever before us have been the *provings of the possibility;* many of us have seen the fillings of *good* "zinc-phosphate" which have done two, three, four, and more years of service, and are yet, practically, as good as when they were made. I have fillings in the mouths of patients, upon the articulating faces of which I punctured little indentations, that after two years of service remain just as they were at first. I have fillings in mouths in which beautiful gold-work failed largely, in from one to two years, and in which zinc-chloride lasts but a few months, that now, after more than six years of trial, are about as good as when introduced.

These results are partly from Poulson's rock-crystal zinc-phosphate—a material which has not been produced of late, and others from carefully made 53° fluid perfected by "natural separation."

It is from the behavior of this material, which has been *tried* and *found worthy*, that I collate the "tests" for a "good" zinc-phosphate, and my experience indicates that just in proportion as any of these preparations respond *approximately* to the tests given, so will fillings made from them *average* durable and satisfactory.

Like the other zinc plastics, the zinc-phosphates are prepared for use in the form of a powder and fluid or crystals.

The powder is made by treating oxide of zinc with nitric acid, evaporating to dryness, calcining, and pulverizing.

This work, so far from being an affair of two or three hours, is a matter of several days and nights, consecutively, for the production of even a moderate amount.

The powder thus prepared is readily distinguishable from the ordinary oxide of zinc powders by its *dry, quick* fineness; a *peculiar* cracking apart of the bulk of powder as the bottle containing it is slowly revolved; by its feeling of *slight grit,* be it never so fine, and by its *decided weight* as compared with any oxide of zinc powder.

This powder, with a *good* fluid or crystal phosphoric acid, is capable of making a very resistant cement.

So far as I am aware, all the acid menstrua are fluid when first made; but, after being prepared, the changes which supervene are peculiar in character and very numerous. In some cases, the limpid fluid or syrup retains its clearness only a few days or weeks, when, gradually, a cloudy appearance is observable; this increases until a thick, whitish paste results; this again separates into sediment and supernatent limpid fluid.

It has been stated that such preparations make "equally good" cements during any and all of these phases; but experience proves that it would be more correct to regard them as all *equally bad*, for no cement made from any such fluid or paste does any credit to "zinc-phosphate."

In some fluids the limpid fluidity changes to turbidity, and, without passing through the paste form, gradually separates into sediment and clear fluid. This fluid "keeps" indefinitely, and, with a proper powder, makes a cement which may appropriately be styled a "bastard" zinc-phosphate. It has a family likeness to the genuine article, but its working, and its comparatively prompt disintegration in the mouth, proves its illegitimacy.

Occasionally a clear fluid is made which, for a time, is *as good as fluid can be*. It maintains its integrity surprisingly; it works quick and well; it makes a reasonably good cement; but it gradually, and without apparent change, except to an expert, deteriorates. This change is merely one from *quick fluidity to slight viscidity*. It is scarcely perceptible except to a practised eye, but is readily noted by almost any one when attention is directed to it. When this change takes place, the working of the cement also changes entirely. It works easier; sets much more slowly; *becomes, generally, much more acceptable*, and of such material *thousands of unreliable* fillings have *already been introduced*.

Up to the present time it has not been PROVEN that any *fluid* makes as good cement as does *crystal* acid. But here, again, is much circumspection necessary; for it seems to be essential to "excellent" results that the syrupy acid should change to hard, rock-like crystals, and *not* to feathery, flaky, waxy crystals.

When rock-like crystals are obtained, they are to be melted in very small quantity, and with very great care, to prevent ebullition, over the flame of a spirit-lamp. It has been directed to use a spoon of platinum for this purpose; but this is not requisite, as any ordinary teaspoon is appropriate.

When melted, the crystals assume a clear, sticky viscidity. This syrup is then to be scraped from the tip of the spoon-bowl — where it is best that it be placed for melting — on to the glass-mixing slab. The powder should *now* be poured out, as thus a moment is given the syrup for cooling. If mixed while the syrup is hot, the zinc-phosphate sets almost instantly, indeed, so rapidly is this change sometimes effected, and with such evolution of heat, that it is dangerous to take the mass in the fingers lest one be burned.

During the past five years some apparently substantial advance has been made in knowledge of zinc-phosphate. As regards the *powder*, it has been definitely settled that the clear nitrated oxide of zinc is as good, if not positively *better*, for the making of "zinc-phosphate" than any mixture has proven to be.

The most promising of all the suggested additions was the *heavy oxide of magnesia;* this salt, in varying proportions, was for a time enthusiastically endorsed, and the results seemed to me, from their evident obduracy, to be very desirable; but time and repeated experimentation proved two things, viz., that great difference existed in the *working qualities* of the *heavy oxide*, some setting quickly with but indifferent hardness, while others set reasonably promptly and with decided hardness.

Again it was found that the *heavy oxide* "time aged," and in so doing lost all its valuable characteristics; and, finally, the statistics of four years showed nothing of positive value as pertaining to even the best of the samples. For these reasons I have finally abandoned this last of the "admixtures" with zinc-phosphate powder.

For the *fluid* I think we can claim decided advance in the so-called *"gelatinizing"* of it. This is done by incorporating gelatin with the usual solution of glacial phosphoric acid. This is used in varying proportions by different makers; but my experience indicates that twenty-five or thirty grains to the fluid ounce of water gives desirable results.

No other modification of fluid has proved so worthy as this, for, although the "gelatinized fluid" has not equaled the record which pertains to the old "rock crystals" or to the "short-lived clear fluid," it nevertheless possesses more *practical value* than either of these, from the fact of its much longer maintenance of integrity.

Rock-crystals and "excellent" clear fluid commence to lose their integrity in from four to six months; after this the zinc-phosphate made from them is less and less durable until it becomes very poor. Gelatinized fluid maintains its integrity— less than "excellent" in degree, but "very satisfactory"—for twelve or sixteen months, and even then makes creditable results. Thus it is that as zinc-phosphate should *never* be used for a *filling material* without the decided statement that it is *experimental*, and, on the contrary, that as it is eminently worthy in its *legitimate utilization*, so gelatinized fluid is regarded by plastic-workers with exceeding favor.

The legitimate uses for zinc-phosphates are lining cavities, strengthening frail walls, largely filling such cavities as are to be filled with gold, on the score of expense; or with amalgam, on the score of easy removal and possible contingencies, or for durability of filling and maintenance of color; and for increased adhesion of fillings in cavities with but slight possibilities of retention, which are to be filled with *combination* fillings of zinc-phosphate and amalgam; by such I mean fillings in which the two materials are introduced while *both* are plastic, and thus the adhesion of the zinc-phosphate and the resistance to attrition of the amalgam are utilized in one filling.*

These purposes zinc-phosphate subserves admirably, and for these it is much more important that we have a "long-lived" fluid than that we have an evanescent menstruum which *may* make a material *insoluble* in the fluids of the mouth.

I have said that "gelatinizing" seems to control the changes which pertain to the phosphorus in marked degree, but it is nevertheless true that *peculiarities* pertain to these fluids in almost equal degree to those of the old methods of manufacture.

Made by like measurements and weights, dissolved by like

* See Appendix, Sec. 7.

heating for given lengths of time, filtered by similar filtration, no two consecutive results are *precisely* the same, and some are widely different.

We have, *first*, those which maintain an almost absolute integrity as amber-colored, syrupy fluids; *second*, those which are whitish, viscid and semi-opaque as the result of the first phosphoric impress, and which maintain this condition remarkably; *third*, those which deposit a stratum of pure white sediment and become a limpid, supernatent syrup; *fourth*, those which deposit a mass of clear crystals upon the bottom of the bottle, and which evince no other change either of color or consistence; *fifth*, those which have dark scum floating upon the surface; *sixth*, those that have dark sediment; and *seventh*, those that divide into strata of varied consistence and color.

But now we find, *by testing*, that these changes — unlike those pertaining to the fluids of former days — not only do not produce that marked deterioration from "excellent" to "worthless," but leave the fluid in such fairly good condition as to detract but slightly from its original worth.

The "maintenance of integrity" of phosphoric menstrua had been practically abandoned, as the worthlessness of the hermetically sealed tubes had been demonstrated, and the assertions that zinc-phosphate preparations worked best for a short time *after the opening* of the package were found to be unreliable, and when it seemed decided that the menstrua *held within themselves the elements for their own destruction*, and it was therefore with much satisfaction that the workings of the modified fluids were noted until such time as years had given the assurance of some reliability. As with the other zinc-phosphate fluids, experiments have demonstrated that whatever changes take place in these menstrua occur as the result of *time*, and that these are *practically the same* whether the fluid be kept in tightly-stoppered bottles or in open bottles, or whether it be *alternately* exposed to, and protected from, both air and light.

It must be understood that the illustrations given of changes and peculiarities of fluid and semi-solid menstrua are merely a typal few of the many which have been recognized, and which are constantly being added to, and that it would be wholly unprofitable to continue citations of them, as practitioners cannot, by this means, be made judges of the comparative value of the various zinc-phosphate and oxy-phosphate preparations.

It is *only* by "working-tests" that this can be decided, and it is for this reason that I shall direct particular attention to them. My facilities for noting these have been very extensive, for I have been in receipt of all the preparations made, both in Europe and our own country, for nearly nine years; of these I had introduced, more than six years ago, over five hundred experimental fillings, all tabulated, and all under frequent inspection, and from this observation I have obtained such data as I have reason to believe will give zinc-phosphate — at least, as now made — its proper value.

It has been intimated that the zinc-plastics seem destined to supersede amalgam; but I regard such opinion as without the slightest foundation, and based alike upon ignorance of the progress which has been made in amalgams, and of the deficiencies which notably characterize the oxy-chlorides, the oxy-phosphates, and even the best of the zinc-phosphates.

It is true that the zinc-plastics have a better color than amalgam, but in no other respect do they compare favorably with it as filling materials.

When a zinc-phosphate is really "good," it is very good; and, although it is much more difficult to work than any other plastic filling material, yet the practice which has, necessarily, to be given to overcome this, is by no means unworthily bestowed, for the durability of a good zinc-phosphate filling is at least from three to five times that of the average oxy-chloride of zinc filling, and its density makes it of great value even on the articulating faces of teeth.* On the contrary, poor zinc-phosphates, and even the best oxy-phosphates, are, *comparatively*, worthless materials. They are better than the average oxy-chlorides, but compared with good zinc-phosphates they are quite inferior. They are, moreover, very alluring and equally deceptive; for they mix nicely, work easily, set slowly, and *apparently* make very nice fillings; but it requires only a few months — six to twelve — to demonstrate that, although they *promise much*, they perform but little.

---

* *Especially* on articulating faces of teeth.

I wrote of them, nearly ten years ago, that they made "a thick, tough, doughy mass; they are easily and handsomely introduced into cavities; they set slowly, and thus give time for careful manipulation; they finish nicely, and they make *miserable, good-for-nothing fillings!*" and the opinion then offered has been most fully corroborated by the results which have since been noted.

Mixing Zinc-phosphate.

In mixing zinc-phosphate *for "testing,"* a portion of fluid equal to one or two drops, or ten grains of crystals, melted, should be placed upon a glass slab, and *more than sufficient* powder should be poured out near to it. A bulk of powder *about equal in size* to the bulk of fluid should then be mixed with the fluid by means of a suitably shaped steel spatula. This should make a mixture of a thick, creamy consistence.

Then a little more powder should be added and *quickly* and *forcibly* made into mass by thorough spatulation.

It is possible that yet *a little more* powder may be required to give the mass the desired consistence for strength and durability, but it is better that the "mix" be a proper one by the *two* additions of powder. The reason for this is, that zinc-phosphate is a quick-setting hydraulic cement, and the chemical combination is one that commences promptly, and should be the least possibly interfered with during its progress by the successive introduction of ingredients.

The mass should be of a *putty-like* consistence, though some varieties are directed to be made "stiff." In "testing," these should be made according to directions, so as to give them the benefit of every possibility.

The mass should then be scraped up on the spatula and taken from it by the thumb and forefinger.*

The *first test* for a "good" zinc-phosphate is, that it requires *decided* force to take it from the spatula.

The mass should then be *kneaded*—NOT ROLLED—by two or three gentle motions of the *thumbs* and *forefingers.* The warmth of these makes the mass *slightly* more *plastic,* and the *kneading* is of *great value* in producing homogeneity.

* See Appendix, Sec. 5.

The fingers should be scrupulously clean, or the material will be soiled.

The mass should then be upon the end of the forefinger of the left hand, from which it should be removed by the spatula and placed on the palm of the same hand, and then gently rolled into a round, pill-like pellet.

> NOTE.— Up to this point the directions for preparing zinc-phosphate for "filling" are the same as for "testing." At this point the pellet should be made oval or elongated for "filling," but should be as spheroidal as possible for "testing," the object being to attain a shape which will give the most accurate rebound

The *second* test is, that it roll into pellet with prompt *cessation of adhesion* to the rolling finger, and that it *does not adhere* to the finger when it is pressed upon it to determine this.

Third test.— In *one* or *two* minutes, by the watch, it should *glaze* and *rebound* when dropped upon wood, glass, porcelain, or marble, but *not necessarily* upon *metal*.

Fourth test.— In *five* minutes, it should give a porcelain-like feel and sound when tapped gently on the edges of the lower teeth, and should have no "sticky" feeling when pressed between the fingers.

Fifth test.— In *ten* or *fifteen* minutes it should be resistant when taken between the teeth and bitten upon; and, if bitten with sufficient force, it should *break* with a clean, sharp fracture.

Sixth test.— In *twenty* minutes it should be difficult of puncture to the point of the spatula, even with considerable force; and it should take a fine and *persistent* burnish.*

Seventh test.— In *thirty* minutes it should have no taste, or at most an astringent, metallic taste, like that of chloride of zinc, but *not acid*.

It should now be impenetrable to the point of the spatula, even with decided force, and should rebound, with ivory-like elasticity, when dropped from no greater height than an inch.

I will repeat, that an experience of eight years decides me in the opinion that just in proportion as a zinc-phosphate cement responds approximately or *accurately* to these tests, so does it make, in a majority of cases, the most durable and the most satisfactory fillings possible to this filling material.

---

* This test is markedly modified by "gelatinized" fluid, in that it is frequently an hour or more before puncture is difficult; but even these pellets eventually become very hard.

If an oxy-phosphate or zinc-phosphate is "poor" or "doubtful," it will respond to "testing" as follows:

First.— It will not require so much force, even when mixed "stiff," to take the mix from the spatula.

Second.— It will adhere to the finger while rolling into pellet, and will be lifted from the palm of the hand by pressing the finger *lightly* upon it.

Third.— In from *one* to *two* minutes it will *not glaze* even though it rebound; and it will, usually, *not rebound*, but will fall "dead" or motionless.

Fourth.— In *five* minutes it *may* give porcelain-like feel and sound when tapped gently upon the edges of the lower teeth, but will usually have a "sticky" feeling if taken and pressed between the fingers.

Fifth.— In *ten* or *fifteen* minutes it will not be resistant between the teeth; but if bitten upon with gentle, sufficient force, will *admit of indentation*.

Sixth.— In *twenty* minutes it will permit of puncture by the point of the spatula, even with moderate force; and it will not take a *persistent* burnish, even though it admit of temporary polish from the burnisher.

Seventh.— In *thirty* minutes it will still have decided taste,— metallic, astringent,— and sometimes even more than this, *decidedly acid*. It will, even yet, permit of penetration by the point of the spatula, and will cut like plaster of Paris, or, even worse, like *wax*.

One of the most marked tests of zinc-phosphate is the prompt loss of taste on the part of a good material — in an hour or less — and the *persistence* with which an inferior material will retain the astringent, and even the *acid* taste, sometimes for days.

This is attributable to the imperfect combination of the menstruum and powder, which results in a mixture of phosphoric acid and oxide or nitrate of zinc — *with acid reaction* — instead of forming the *neutral* hydraulic cement recognized by plastic dentistry as "ZINC-PHOSPHATE."

DIRECTIONS FOR USING ZINC-PHOSPHATE.

It has been noted that the "mixing" of zinc-phosphate is the same for "testing" as it is for use as a filling material, up to the point when the mass is rolled into pellet on the palm of the hand. For filling, this pellet should be made oval or elongated, to further its facile introduction to the cavity of decay.

The pellet should then be taken by a pair of foil-pliers and placed in the prepared cavity.

The filling should be introduced by appropriate *smooth*, flat, or round-ended instruments, which may be found in Nos. 2, 3, 5, 6, 7, and 8 of the set of amalgam filling instruments.

The filling should be worked into place, and the *condensation of the material* effected by pressure of the face of the filling in precisely the same manner as is done with soft foil, with the exception that the same *smooth-faced* instruments should be used instead of pointed or serrated fillers.

It will be found advantageous to touch the faces of the instruments to an "oil-pad,"— see Gutta-percha,— as this prevents adhesion to them of the filling material; particularly is this useful in filling approximal cavities and in "facing" built-up crowns.

The superfluous portions of the mass — should there be any — ought to be so manipulated as that they will overhang the **edges of the cavity.** When this is done, they readily break away, leaving the filling nicely adapted to the walls of the cavity, and it also secures a *smooth, compressed surface* to the filling.

In filling large, difficult, and, in a measure, inaccessible cavities, it is essential to good results that *several mixings* should be made, *each of the various portions being introduced separately, and manipulated just like a pellet of soft foil.*

In cavities between teeth which extend up to, and particularly *under*, the gum, it is necessary to fill out to the cervical edge and downward for at least *one-third* of the cavity with gutta-percha, as a double preventive to decay of the tooth and disintegration of the zinc-phosphate.

NOTE.—The disintegration of oxy-chloride of zinc at the cervical portion of the filling has been referred to; and it will therefore only be necessary to mention that zinc-phosphate has the same peculiarity, but in less degree. It is, however, sufficiently vulnerable at that point to render it imperative that a "guard" of gutta-percha or submarine amalgam, according as the tooth is more or less conspicuous, should be first introduced.

After the introduction of the filling, it should be kept dry for at least *five minutes*. It should then be waxed or varnished, that moisture may be promptly and effectually excluded.

 ℞.—White Wax . . . . 1 part.
   Dark, unrefined Resin . . . . 5 parts.

Melt and stir thoroughly together. Work it and make it into sticks, using powdered soapstone to prevent adhesion to the fingers. This wax is used by heating instrument No. 7, and taking a small portion and placing it, molten, upon the filling. By reheating the instrument the surface of the filling can usually be nicely covered with a thin coating of wax, which hardens almost immediately and adheres with much tenacity. Upon articulating fillings it is better to spread the wax by heating, quite hot, No. 2, ball burnisher, after having placed the wax in position by No. 7. For cleaning the instruments after using wax, they should be heated and wiped off with paper. This is an easy and satisfactory method of accomplishing a somewhat troublesome task. If varnish be used it should be allowed to dry thoroughly.

 ℞.—Virgin Rubber . . . . grs. v.
   Chloroform . . . . f℥ij.

Dissolve thoroughly by succussion. This takes one or two weeks to accomplish.

   Gum Mastich . . . . ℈ij.
   Chloroform . . . . f℥i.

Dissolve.—This dissolves with ease. Mix the two and shake well.

NOTE.—This varnish—the suggestion of Dr. C. F. Ives—is so excellent that I deem it unwise to offer any other. Its general adoption will prove eminently advantageous, as it is not only a "protector" but a "liner" and "non-conductor," while experiments—made this year by Mr. Henry A. Mansfield—have proved it to be the only "intermediate" among all the many varnishes used by dentists.*

---

 * An "intermediate" prevents *any* deleterious impress from passing it. Used for this purpose the varnish should be placed in position upon fine muslin.

If it is not easily practicable to maintain the five minutes of dryness suggested for the filling, this should be waxed or varnished immediately upon completion.

If any further finishing—for articulation, etc.—is required the filling should be allowed to harden for at least *thirty minutes* —longer time is better, if possible—before this is done; *after which it should again be covered as before.*

In finishing "zinc-phosphate" fillings, cutting instruments — like No. 9, for example — files, burrs, stones, and disks are admissible. When files or burrs are used, they should be either *perfectly* dry or *thoroughly* wet; as these conditions, in great degree, prevent them from becoming clogged.

The face of the filling is better if burnished after it is thoroughly hardened. For this purpose a steel or agate burnisher can be used. If the face of the filling is marked from using a steel burnisher, it is readily cleaned with pulverized pumice on a soft pine stick.

It seems proper that a few final directions, or rather *suggestions*, should be given with a view to completeness rather than importance.

In taking the "crystals" from the bottle, it is far better that sufficient be scraped up and shaken or turned out into a spoon, than that the contents of the bottle be warmed by placing it in hot water; as *every such heating is eminently detrimental to a good crystal.* Ordinary crystals it does not appear to injure much, if any; for they continue to make cement about as at first.*

It has been directed that "more than sufficient" powder should be poured out preparatory to mixing the mass. The reason for this is, that even a respectably good zinc-phosphate will "set" so promptly as to require that no time be *needlessly* lost in its preparation and insertion, which would necessarily be the case if one were obliged to stop and replenish the stock of powder.

In returning the *surplus* powder after filling, care is requisite not to return any which has been touched by the fluid.

It is sufficient to say that *no menstruum*, either crystal or fluid, should be used after it ceases to make a cement which will respond to the "tests" for goodness.

---

*The taking of "fluid" is best and most economically done by means of a small piece of stick.

The best method for removing from the fingers such portions of zinc-phosphate as may adhere to them, is by rubbing them, *while wet*, with the edge of the glass mixing-slab.

For the removal of adherent material from the slab, the spatula will prove convenient and effective — scraping the slab while holding it in a stream of water.

In *removing* and *replacing* the corks from both menstruum and powder bottles, it is better to twist them slowly and steadily, *always to the right*, as this will prevent breaking them.

Conclusions. — In estimating the value of zinc-phosphate cements, I have realized that much discount must be made upon the statements of most of those who manufacture and dispense them. They have already been invested with such names as indicate gradations of obduracy from porcelain to adamant, but I do not think the provings will rank any such appellations as appropriate, or, indeed, as anything but deceptive.

Zinc-phosphate cannot be regarded as an "insoluble" cement, but must be accepted as "not *very* soluble" in the fluids of the mouth. It can, however, be graded as much better than oxychloride of zinc.

It can, *in no wise*, be esteemed as even a moderate approach to the "ideal filling material," for, although it possesses many of the desired attributes, it is, nevertheless, deficient in a sufficient number of them to place it entirely without the pale of consideration in this regard. It has not only been suggested, but strongly recommended as a "pulp-capper," and this merely because it does not irritate that organ; but it must be remembered that a material may be non-irritant and yet be eminently capable of pulp devitalization — *vide* Cadmium; therefore, I have always warned my classes against this use of the material until it has been *proven*, by careful, patient experimentation, to be worthy of reliance in this particular, and up to this time my own convictions are rather adverse than favorable to its claims.

It seems to be especially adapted for articulating, approximal — under guarding — and labial cavities in *frail*, *pulpless* teeth; or in teeth containing vital pulps, with these protected by temporary stopping or rubber varnish.

The results of ten years place both oxy-phosphate and nitro-phosphate as "excellent" liners, strengtheners of frail walls, and solid sub-strata for external reliable filling materials.*

It will now be seen that from the plastic standpoint, while there exists no such use for zinc-phosphate as would warrant any unduly considerable demand, it is, nevertheless, an important and satisfactory adjunct, and one that will meet the requirements of certain cases, which, without it, would be almost insurmountable.

ARTICLE XVII.

TEMPORARY STOPPING.

A MATERIAL for this purpose has been suggested to me, which I deem worthy of especial notice. It is a mixture of red gutta-percha base-plate, white wax, silex, and feldspar.

In a sand-bath, over a spirit-lamp, place a small glass beaker or other convenient vessel. In this melt two pennyweights of white wax. When this is melted, add to it six pennyweights of gutta-percha base-plate, cut into small pieces; when this is melted, add three pennyweights of finely pulverized silex and three pennyweights of powdered feldspar. Stir all thoroughly together and let the mixture partially cool. When the mass becomes doughy, press it between two plates of glass, first wetting them to prevent adhesion. This compound fulfils the requirements of a "temporary stopping" very acceptably.

As at present modified this material has taken, fully, the important position among "plastics" which seemed to render it "worthy of especial notice" in the first edition. Indeed, the uses for it are, at present, so varied that its name seems almost a misnomer. So far from being only a "temporary stopping" it is found worthy of the greatest reliance as a *permanent stopping* for portions of canals and for bulbous parts of pulp cavities.

It has proven to be the only compound of gutta-percha which —though not absolutely "tight"—is graded as "non-leaky," and therefore ranks as one of our *only two* "intermediates" and as one of our *only two* reliable coverings for arsenical applications.

*See Appendix, Sec. 7.

TEMPORARY STOPPING.

As now made it is not only a more resistant temporary stopping, but even does duty as retaining wedges for spaces between teeth. It will thus be recognized as an "indispensable" to the practice of dentistry.

℞.—White Wax . . . . 1 dwt. (full).
Red Gutta-Percha Base-Plate . . . . 4 dwt.
Precipitated Chalk . . . . 4 dwt.

In a small, porcelain covered, iron ladle melt the wax. When melted add to it the gutta-percha base-plate, cut into small pieces; this must be *carefully* and *thoroughly* melted into a smooth thick paste; then add the chalk, and work all together by means of a pestle.

When mixed, take a convenient sized portion and roll it into a ball between the palms of the hands; then placing the ball upon a smooth surface roll it gently into a stick by the fingers of an open hand. Smoothly round the stick by rolling it with a flat piece of wood or plate of glass.

White temporary stopping is both made and used, and I cannot too strongly condemn this ill-advised change of color. As a *pink* material, temporary stopping is a constant warning of danger — if danger exist; a constant guard against leaving any unfinished work; a constant indication of conditions, when reached — as in relief taps, approaches to pulp cavities, etc.; in short, a constant reminder of its presence in all its variety of utilization.

As a *white* material all this advantage is not only lost, but in its similarity to usual gutta-percha stopping it becomes a constant source of possible danger, possible neglect and probable trouble.

As a retaining wedge in front spaces, low heat gutta-percha is even better than temporary stopping, and this, when removed, shows the *pink* temporary stopping as *temporarily* filling the *cavity* or *cavities* beneath. .

Temporary stopping may be warmed either over the flame of a spirit-lamp, while held by a small probe, or upon the upper plate of the G. P. warmer, from which it is taken by a probe, and should be worked with instruments which are *slightly warmed*, if it is to be used as an "intermediate," though this is not necessary when it is used for other purposes.

ARTICLE XVIII.

TECHNICALITIES.

AS in the progress of the development of plastic-filling materials, instruments for manipulation, apparatus for testing and experimentation, and methods for utilization have been devised, and substances—metallic, organic, and inorganic—have been brought into service, so has it been found needful to coin a technical phraseology by which explanations could be made; comparison of views be indulged in; methods referred to; and advantages and disadvantages of operations be discussed, concisely, definitely, and intelligibly.

So strangely apathetic have been the reputed leaders in dentistry, that, notwithstanding the constant reference to these terms which has necessarily been made in practice, as repeated demonstration of power to cope with things impossible of accomplishment by any previous materials and methods has been given, and although attention has been directed to them by public and careful enunciation, they have, up to this present, remained terms possessed of meaning to but few, and these, with rare exceptions, from among the younger and rising members of our profession.

It is noted in journal communications that an occasional use of plastic phraseology is indicative of commencing acceptation, and it is now, beyond question, a matter of immediate import that those who purpose *saving the teeth* of the coming generation, should familiarize themselves with the technicalities of that practice which alone has demonstrated a reasonable capability for the attainment of this, much to be desired, result.

Ageing.— This term is applied to certain changes which occur to almost every variety of every plastic-filling material. It is also used to express the attainment, by any means, of results which are given by time, either desirable or undesirable; as, for instance, the "ageing" of gutta-percha is recognized as detrimental, in that the material becomes brash, losing its toughness; it becomes easily disintegrated by attrition and by the fluids of the mouth; it is crumbly — indisposed to cohere; it is "time-rotted." Oxy-chloride of zinc powder loses its sharpness of response in setting and also its density and hardness, becoming slow-setting, porous, and soft.

The fluids and syrups of oxy-phosphates, and the fluids, syrups, pastes, and crystals of zinc-phosphates, as a rule, lose their value entirely from "ageing," and those which do not, have equally, as a rule, but modified value at any time when compared with an "excellent" menstruum.

Amalgam alloys, on the contrary, after being cut or filed, improve notably from "ageing." Amalgams made from really good alloys are, without exception, very much better after these have been filed up for several months. Ordinary alloys work better when "aged" than they do when "fresh," but the better the alloy the more marked is the improvement due to "ageing." They work much more smoothly, require less mercury, set more slowly and yet with sufficient celerity, and, under every "test," become more desirable.

Reference has been made to the fact that, in considerable degree, this effect of time upon cut or filed alloy may be produced by continued attrition of particles, or by repeated exposure of particles to the air, or, most probably, by a combination of both. This is done by placing the filings in a cylinder of glass, and causing it to revolve with *moderate* speed for the space of several hours, and the process is called "ageing the filings."*

Bulging.—By this is meant a change of shape which takes place in amalgam, and by which the form of a filling is so altered, from spheroiding, as to present a modified convexity of face. This result is attributed to the spheroidal control of the mercury, and is more pronounced as the metals of which the alloy is composed are more yielding in texture, less obdurate in melting, and more deliberate in cooling. The concomitant of *bulging* is *crevicing*, as, in the attainment of a spheroidal form, the filling material is, necessarily, drawn from such cavity-edges as have had to be left straight instead of curved—concave.

This term is also applied to the softening and swelling out which sometimes, though very rarely, occurs in fillings of gutta-percha—notably *red* gutta-percha, or *base-plate*. This result is generally attributed to inferiority of material; but, though

* See Appendix, Sec. 1.

this may be usually a correct ascribing, I am inclined to the opinion that, in some mouths, it will occur with the very best gutta-percha.

Buffering.—This word is used to denote an operation by means of which a yielding material, like gutta-percha, is protected from attrition by a more or less resistant material, like zinc-phosphate or amalgam. It signifies protection from blows or frictional impinging, and is the method by which the plastic-filler is enabled to utilize the tooth-saving power of gutta-percha in frail bicuspids and molars in which are cavities of decay extending up to and under the gum, and opening through upon the articulating faces of the teeth. The "buffer" is generally trunnioned either in dentine or zinc-chloride, though sometimes an acceptable retaining shape can be given to the articulating cavity-edge.

Capping.—As used with reference to *filling*, this word implies completely outside work, in contradistinction to the completely inside work of *pulp-capping*. Whereas, the latter is done with the softest and most yielding materials used in plastic dentistry, the former is done only with those possessing markedly the characteristic of resistance to attrition. It is used to express the thin covering of a yielding material with which a "tap" for future "venting" may be filled. This is usually gutta-percha, a material which must be protected if the "tap" has to be made upon any surface exposed to attrition. The *capping* is done either with zinc-phosphate or amalgam, and if with the latter, "facing" amalgam is usually employed, as, by its marked retention of color, the position of the "tap" is indicated.

Cold-Soldering.—This phrase is used to express one of the most useful possibilities in connection with plastic filling. *It is a property of amalgam, that additions of this material can be secured firmly and homogeneously to either gold or amalgam fillings, whether they be old or of recent introduction.* In this way old fillings are joined on to and made subservient for retaining purposes, in cases of new decay—shallow cavities—encroaching upon or approaching near to such fillings. Large reparations are made with perfect facility in cases where masses of tooth-structure have been broken away from heavily-filled teeth. Fillings are made to present the appearance of, and to be, practically, one large filling, by merely removing the carious tissue from between them and joining them with new filling material — amalgam.

Pins and tubes for pivoting are fixed in position by drilling oval drill-holes in amalgam root-fillings, and securing the pins or tubes with freshly-made amalgam.

Gold bands are held firmly upon crownless roots by filling the canals with amalgam and cold-soldering the bands by building up amalgam crowns within them.

Linings of amalgam are placed in cavities of *doubtful* teeth, which are thus secured against decay, and, after many days, if results are unfavorable, this thin stratum of filling is readily removed. If, on the contrary, the result is favorable, the remaining portion of filling is just as readily added on.

Crowns are partially built up, and if circumstances — such as want of time for completion of operation; urgent demand for services by other patients, etc.,— compel cessation of work, it is a matter of no moment so far as any complication is concerned, for, at any time, the work may be resumed just as if it had not been interrupted. All these, and many other requirements equally convenient to meet and desirable to avail one's self of, are boons bestowed alike upon patient and operator by "cold-soldering."

As in all soldering the face of the metal to be soldered upon must be made bright either by scraping with an excavator, smoothing with a file, or cutting with a burr-drill; a small portion of amalgam is then mixed *very soft*, and, being placed upon the brightened surface, is rubbed over it until it is thoroughly "amalgamated," after which all superfluous amalgam should be removed. Having thus obtained a soldering surface, the appropriate amalgam — indicated by the position of the cavity or the exigencies of the case — should be mixed as usual, and the filling made. The necessary caution in regard to care in using, if such is required, should not be forgotten, and, with such admonition, the patient can be assured that upon the complete hardening of the filling material a result will have been attained which, in strength, durability, and homogeneity, will be precisely the same as though the filling had been originally made with its present contour and dimensions.

Crevicing.— This word is used to express that result referred to under "bulging," when an amalgam filling has assumed such shape, from tendency to spheroid, as to have drawn from the margins of the cavity. It is probable that more amalgam fillings have been unjustly condemned, as worthless, from this "crevicing" than from any other cause. So far from being a serious and irremediable defect, it is *most frequently* very easily and quickly removed, and, by its removal, a filling of great value is secured.

The removal is effected either by filing, stoning, or burring off the superfluous amalgam; and a filling is thus again brought into contour with the cavity edges, which, having bulged in one direction — outwardly — has likewise, from the same cause, become even more accurately than ever in apposition with the cavity walls. Besides this, the filling is usually an old one, or, at least, one of several years' duration, and, during all the time, a formation of sulphides has been progressing, which, by impregnation of contiguous dentine, has rendered that tissue of most congenial potentiality with the surface of the filling. All this harmonious condition of things has been reached by that gentle gradation of change which the experience of many years proves to be productive of extraordinary permanency.

It may, therefore, well be questioned as to whether the removal of such fillings is not alone of doubtful benefit; but, far more than this, acts of *positive wrong* to patients — work which it behooves every conscientious practitioner to weigh well before continuing in the stereotyped condemnation based upon convictions which time has shown to have been founded upon prejudice alone.

If *crevicing* is of such depth as to render the removal of a superficial stratum of amalgam insufficient to restore to the filling a desirable perfection, it is proper that a groove or canal should be cut, with an oval burr or fissure-drill, throughout the length of the crevice. This groove should then be filled with freshly made amalgam, which, as has been stated, will unite homogeneously with the former filling.

It is frequently possible, at the present time, to very much improve the appearance of these creviced amalgam fillings by cutting out a thin stratum of the old amalgam, and replacing it with one of the good color keepers, either "contour" or "contour" and "facing," half and half, or even, in rare cases, where edge-strength is of no special consideration, with "facing" amalgam alone.

Domeing.—This operation consists in building up a pointed or conical dome of gutta-percha from the pulp-cavity towards that portion of the periphery of a filling at which it would be proper to "tap" if crown-tissue were existent. Having thus formed this "dome," the filling, of amalgam, is made, and the point of the dome covered in. If at any future time it becomes advisable to enter canals, the entrance is easily made by drilling through the thin covering of amalgam, and removing the gutta-percha by means of a warmed instrument or small oval burr. The entrance having been effected, and the desired relief having been given, the matrix left by the removal of the *dome* is readily refilled, and the position of the tap made perfectly apparent by capping with facing amalgam.

For superior molars, the dome should point towards the centre of the articulating faces; for inferior molars, the dome should point mesio-*buccally*. Molars are the only teeth which it is ever necessary to dome.

Facing.—In "facing," we have *the* æsthetic of plastic dentistry. It is, in the direction of beauty, that which cold soldering is in the direction of utility. Facing is done with gutta-percha, zinc-phosphate, and amalgam; the incisors and cuspids either with gutta-percha or zinc-phosphate, the bicuspids and molars usually with amalgam, exceptionally with gutta-percha.

Facing is of two grades,—the facing of fillings and the facing of "built-up" teeth. The first is of comparatively circumscribed extent, and is done in cases where decay has progressed in such manner as to impinge upon the labial faces of incisors and cuspids. Cavities in these teeth having been properly lined, if necessary, and filled in contour with contour amalgam, the filling is allowed to harden for about thirty minutes, after which it is *cut into* with the trimmer from the front. The excavation is bounded by the remaining labial enamel on the one side and by the *contour-line* of the filling on the other. The slot thus cut out is, in this way, made of the exact shape of the missing portion of enamel, and the slotted filling is then allowed to harden thoroughly — an hour or more.

If it is desirable to avoid retaining the patient longer at this sitting, the slot may be filled temporarily with low-heat gutta-percha. This is not sufficiently resistant to endanger the frail, bevelled edges of the newly-made amalgam on the contour edge. The slot should be finally filled either with "high heat" white gutta-percha, or with "medium heat" shaded gutta-percha, or with zinc-phosphate, by making the shade desirable, or with "facing" or "front tooth" amalgam. Both gutta-percha and zinc-phosphate may be shaded *yellowish* by a *little* yellow ochre, and *bluish* by a VERY *little* lampblack.

These gutta-percha facings can readily be made to match the color of the teeth so closely as not to attract the least attention. They are quite durable, lasting frequently for several years, and are renewed at any time, if required, in a few minutes.

The zinc-phosphate facings are, as yet, more experimental, but I have a goodly number which, during the past eight years, have needed but infrequent renewal. I never use a zinc-phosphate for facing unless it responds well to all the tests given for "good" material, and I am very careful to maintain dryness during filling and *setting*,— usually with rubber-dam,— and to cover thoroughly with adhesive wax.

In facing "built-up" crowns, it is first *essential* that the crown be built *perfectly* with contour amalgam. In no other way can the contour-lines for facing be attained with any degree of accuracy or beauty. In the first building the whole anatomy and expression of the crown can be given, and the proper articulation is secured. The result is a crown which, for size, appearance, strength, and utility, would be all that could be desired; but the insuperable objection to such a crown is, that it will probably change to a color worse even than that of gold. To obviate this, the buccal face of the crown is concaved most delicately, most accurately, and artistically. This is no easy task; for the slightest slip of the trimmer or excavator, and, above all, the least *mal à propos* whirl of the burr-drill, will mar the beauty of the contour-lines.

The concavity being made, if it is to be filled with facing amalgam,— tin, silver, and zinc,— the work can be well done immediately, and, for a vast majority of cases, I prefer this facing to any other, as the built-up crowns appertain, almost exclusively, to *bicuspids* and *molars*, which, with such facings, almost always make very presentable and, indeed, pretty operations.

But it is nevertheless true that, *occasionally*, facing amalgam will discolor, and it is for this reason that we sometimes have to resort to the gutta-percha and zinc-phosphate facings for bicuspids. When this is so, the concavities should be made *much deeper* then is needed for the using of amalgam, and mechanical retaining-slots, grooves, or undercuts should be made for the securing of the facing.

In these cases gutta-percha has made a good record, usually doing service for from two to five years, and then requiring but a thin film of additional facing.

Much thought, labor, and experimentation have been bestowed upon the accomplishment of this *desideratum* in plastic work, and the results have been such as to cause the plastic-fillers to feel well compensated for all their toil.

But the satisfaction of the workers has been a slight thing in comparison with the outpourings of delight and gratitude which have come from patients, especially lady patients, who have seen their frail but beautiful "pearls" crumble away from "the jewels of gold," until, while yet in young life, they had come to smile smiles of barbaric magnificence as the "ruby portals" revealed a terrible preponderance of the brilliant triumphs of "first-class" dentistry.

These are the dentures which, taken after the dropping out of the third or fourth line of such work, after the expenditure of dollars by the hundred until the hundreds had wellnigh reached *a thousand*, after the remnants had come to be a jagged row of roots and broken cusps and edges, have been so "built up" and "faced" as to have in such degree restored the original beauty of expression to their possessors, that words seemed inadequate to do their feelings justice, and *written praises* have blessed the day on which they learned of plastic dentistry.

Frotting.— This term is from the French, "frotter," and signifies "to rub." It is applied to the method by which certain

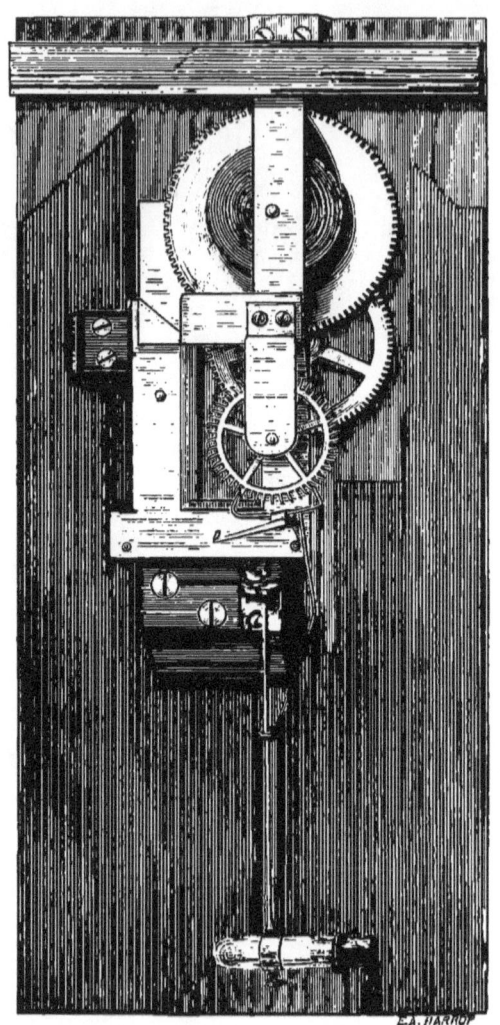

kinds of plastic materials are tested for "probable wear." These are oxy-chlorides, oxy-phosphates, and zinc-phosphates. The illustration of the machine for "frotting" will give so clear an idea of its working as to need no description, and it will therefore suffice for me to say that pellets of the materials to be tested are made into pill form and placed in a small tube containing water, slightly acid, slightly alkaline, or neutral, and are then frotted by the pendulum-like motion given to the tube by wheels arranged like clock-works and driven by a powerful spring.

In this manner the relative durability of such plastics as disintegrate in the fluids of the mouth is quickly tested, and quite a number of materials, which, from the enthusiastic and emphatic assertions of their makers, would have given rise to costly and tedious lines of disappointing experimentation, have been strangled almost at their birth.

Guarding.— For many years it has been noted that fillings are prone to fail at the cervical edge. So frequently is this the case, that this part of the cavity wall has come to be known as "the vulnerable spot." In foil practice, this failure has ever been attributed to "defective manipulation;" but this reasoning has always been a matter for peculiar comment upon the part of the plastic-fillers, and the explanation has been regarded by them as not at all satisfactory. They have thought it strange that the easiest' part of the foil-filler's work — the introduction of the foundation or first pieces — should have been so universally *badly done*. They have thought that, of all the filling, the cervical portion would be the best, and that, in consequence, it should the least permit recurrence of decay. They have noted the same unsatisfactory results at the cervical margins of cavities filled with plastic fillings, and they *could not* ascribe these failures to "defective manipulation;" for they *knew* the manipulation of the filling material at this point was as perfect as it possibly could be. They *knew* that there was no "loosening of the first pieces;" no "slipping of the first layer;" no "leakage from turning;" no "crumbling of cavity walls from malleting;" and yet they knew that the *vulnerable spot* existed for them just the same as it did for the "defective manipulators."

It has also been observed that by far the most frequent and most extensive cavities of recurring decay are found in connection with gold-fillings; that *second* in relative proportion, numerically, the cervical edge of cavities is vulnerable under tin and silver amalgams; that it is vulnerable in *third* degree under fillings of tin-foil, or of silver, tin, and copper amalgams; that it is still less vulnerable — in *fourth* degree — under gutta-percha fillings; and that here the grades of vulnerability of cervical tooth-bone cease.

And now, it has been noted that, although recurrence of cavity formation continues, it does so at the expense of filling materials instead of tooth-bone. The fillings of oxy-chloride of zinc, oxy-phosphate of zinc, and of zinc-phosphate, in the order given, disintegrate at *their* cervical margins; the first, notably; the second quite promptly, and the third, though much more durable, far too quickly.

To the mind of a plastic-filler the "defective manipulation" theory offers no explanation whatever of all this; but with the Palmer theory of *relative consonance of potential*, or "compatibility between filling material and tooth-bone," everything seems clear and easy of explanation.

It is understood, by consent, that all substances have a definite molecular motion which is expressed by the word "potential." It is proven by galvanometric experiments that contact between almost any two substances — using the word in its general sense — develops electrical phenomena.

It is agreed that needle deflection denotes difference in potential between substances just in proportion to the magnitude of the deflection.

It is agreed that in proportion to difference in potential so will contact of substances produce electrical and chemical results.

It is agreed that one of the substances becomes *positive* to the other, and that the other becomes *negative* to the first.

It is agreed that, under the action excited by contact, the substance which is proven to be *positive* is markedly disintegrated, and it is inversely deduced that disintegration of substance proves *positivity* to that substance.

It is agreed that *moisture* increases most notably the rapidity and extent of this disintegration; and, that moisture pos sessed of certain characteristics which render it *neutral*—as it is termed—least facilitates this action; while moisture possessed of other characteristics—acidity, alkalinity, etc., — most promotes it.

Thus far all is clearly scientific, and the plastic-fillers accept this as a *basis* upon which to offer an explanation.

They find aid in this matter from observation of filling materials in their relation to the tables of "conductivity;" from their relative possession of the characteristic of "liability to tarnish,"—formation of more or less soluble salts; from their possession of physical integrity, or, liability to disintegration by chemical action ; and from their relative capability for easy, and consequently more perfect, adaptation to cavity walls.

All these considerations afford subject-matter for thought to the plastic-filler, while the gold-filler, having but one material to think upon, can think of but one thing as promising success, viz., perfection in manipulation ; and but one reason can present itself to his mind as explanatory of failure, which is "defective manipulation."

With thought upon the points which I have presented, there has come an inquiry into the varied characteristics of all the plastic materials that are subservient to the *saving of teeth;* a systematic investigation of their behavior under all possible conditions, and in all possible positions, in the mouth ; a practical testing of the harmonious and inharmonious relations existing between theory and experiment; and it is with peculiar satisfaction that I am able to assert the singular harmony between theory and practice which has been proven to exist by the tabulated work of the past twenty years.

So happily successful has been the meeting of indications by means based upon the principles of the "compatibility" theory, that the most desperate cases, in the way of large numbers of dreadfully inaccessible cavities filled with the most sensitive tissue, and in teeth of the softest and frailest structure, are regarded by the educated plastic-fillers with perfect equanimity, and are so treated and filled as to prove eminently satisfactory alike to patient and operator.

Among the most convincing proofs of the correctness of the theory upon which the plastic-fillers base their choice of material for saving teeth, is the operation under discussion, known as "guarding."

"Guarding" is the placing of a material in apposition with the cervical wall of a cavity of decay, which shall, by its possession of certain physical characteristics, act under certain "law" to prevent, in greatest degree, the recurrence of decay at that "vulnerable spot."

It is to the plastic-fillers of the present day that dentistry is indebted for the utilization of tin-foil as a *guard* at this point. It is true that tin-foil has long been used by the most prominent gold-fillers as a means for the "better saving" of soft teeth, but its advocacy is found, *by reference to the record*, to be based upon the *greater softness* of the material, and its consequent *easier and more perfect adaptation* to the cavity walls.

Thus it will be seen that the *manipulative idea* was ever so prominent as to be the only thing thought worthy of attention.

The further proof that this is so may be found in the constant repetition, upon scores of occasions, that this *soft* material was only intended to subserve a *temporary* purpose; for it is everywhere, and at all times, agreed that such fillings were to be allowed to remain only until the teeth become thoroughly calcified, when *they were to be removed and* PERMANENT *fillings of* GOLD *were to be introduced!*

The gold work upon soft teeth has been founded on this principle, and has been practised under this teaching for the past fifty years, and I regard it as unnecessary that I should waste one word in reference to its *record*.

It is now over twenty-five years since I commenced the use of tin-foil, amalgam, and gutta-percha as guard-fillings at the cervical walls; *not* because they were *soft* and could be *more perfectly adapted* to the parietes, but, EMPIRICALLY, because I had noted that fillings made from these materials permitted occurrence of decay less promptly than did gold, even when worked by the best manipulators of that day; not with the avowed intention of removing them when the tooth-structure became thoroughly calcified, but with the intention of *renewing* them when decay should eventually recur, which I knew would probably be the case at some future time, even though the recurrence should be much retarded.

This, which at that time was *empiricism,* has become, by the adoption of the Palmer theory, the strictest following of *science.* Guarding by gutta-percha or amalgam is now done with full reference to the known requirements, the physical capabilities for response to these upon the part of the filling material, and a knowledge of future probabilities based upon definite data which, in turn, rest upon definite, acknowledged "law."

In cavities which are accessible, in which dryness can be attained and maintained, in which lining with oxy-chloride and final filling with either a single or a combination amalgam is thought to be the proper practice, a *guard* of gutta-percha is the thing indicated. This should be made as thin as possible consistent with certainty that it thoroughly protects the cervical edge. It may be given bulk in moderate degree within the cavity, and worked to a feather-edge at the cervical margin. It is better that it protrude a little, so that the lining and final filling having been accomplished, and the amalgam having sufficiently set, the "guard" may be neatly trimmed off with a heated instrument. This insures nice adaptation, desirable finish, and excellent protection.

In cavities which are inaccessible and which run under the gum, and in which it would be impossible to maintain dryness except by very inflictive methods,—dams, dam-clamps, or dam-ligatures,—it is far better to trust to submarine amalgam. This should be mixed reasonably plastic, sufficiently so for perfectly easy manipulation, and, having attained what dryness may be possible, it should be carefully tapped into complete apposition with the wall. It should then be wafered, that firmer consistence may be given the material. It should be built up decidedly thicker than a guard of gutta-percha, that, when set, it may be trimmed in such fashion as to leave a *neck portion*, which may be utilized, if desired, for the maintenance of dryness, by rubber-dam, during the further lining or filling of the remaining portion of the cavity.

When fillings of gold fail at these "vulnerable spots," *guarding* with plastics is a very comfortable and satisfactory method of repairing damages. The guarding, even in this comparatively inaccessible position,—a work which, with gold, would be increasedly difficult and painful,—is almost invariably less painful, less tedious, less expensive, and more permanent than was the original operation.

When fillings of gutta-percha fail at cervical edges, they may readily be repaired either by adding gutta-percha, which is done by first softening the edge of the original filling by heated instruments and then joining the required amount of material to it, or by grooving the cervical margin and filling with submarine amalgam.

If amalgam fillings fail at cervical edges, it is only needed, after the required excavating, that a guard of "submarine" shall be cold-soldered to the old filling.

As conclusion to "guarding," I would say that the plastic school of dentistry holds to the view that the "vulnerable spot" is found to be located at the cervical edge, not from any unusual or insurmountable difficulty which precludes the possibility of perfect manipulation of material at this point, but because in all cases, and particularly in soft-tissued teeth containing vital pulps, *moisture* is here soonest, most constantly, and most abundantly brought in contact with the filling material. It holds that this moisture is not alone *that* from the outside, and which by *leakage*—as this term is generally understood— might notably aid "defective manipulation" in its work of destruction; but that it is also *that* from the inside, the fluid which is essential to the maintenance of vitality, and which is thus ever present in living tissue.

It holds that, as this moisture is gradually brought into contact with filling material, a degree of electro-chemical disintegration of the *positive* substance occurs, which, in rapidity and extent, is in direct ratio with the existing difference in potential between the two substances in contact.

It holds that the difference in potential between gold and tooth-bone is proven to be very great, and the *positivity* of tooth-bone to gold is equally proven, by the rapid recurrence of decay at this cervical margin, when cavities in teeth of markedly soft structure are filled with gold.

It holds that the lessening difference in potential between amalgam and tooth-bone and gutta-percha and tooth-bone is proven by that comparative immunity from decay which, in direct ratio, is found to result when cavities in soft teeth are filled with, or guarded by, these materials.

It holds that the *negativity* of tooth-bone to filling material is proven by the disintegration of the fillings made with zinc plastics; and thus, while it concedes to these plastics a tooth-saving power, in a certain sense, it nevertheless denies to them the ability to perform "guard" duty.

Heating.—This term is applied to various processes and results; but its exact signification in any given case is understood from the connections in which it is used; thus the "heating" of gutta-percha refers to its softening preparatory to using it for filling, and is understood to be a process which *must* be done with much care for such gutta-perchas as require less than 210° F. for proper softening; and equally *must* be done, and with great care, upon metal plates subjected to the direct flame of gas or the spirit-lamp if "high heat" gutta-percha is to be employed.

The "heating" of instruments is understood to refer to the preparing of these for the insertion or finishing of gutta-percha fillings of all the various grades, a process which is accomplished by a tool-heater or, less conveniently, by heating them in the flame of a spirit-lamp.

The "heating" of gutta-percha for testing its grade, and its relative quality as pertaining to any given grade, is a process which is conducted on a gutta-percha tester, and which gives the exact thermometric grading of different samples of this filling material.

The "heating" of zinc-phosphate menstruum refers to the melting of the crystals, a process which requires much care that it be done without permitting the boiling of the syrupy fluid which results. This "heating" is done upon a small spoon of platinum or silver, which is held *high up* over the flame of a spirit-lamp, that the melting may be done gradually, and may be kept perfectly under control.

The "heating" of a zinc-phosphate mix is that generation of heat which always accompanies the union of the menstruum and powder. It is of very varied intensity and duration; but if of decided intensity,— sometimes sufficient to burn,— or of decided length of duration, it is a fair indication of questionable material. The heating of a good zinc-phosphate, though perceptible, is neither intense nor of long duration.

Lining.— This operation, as its name implies, consists in covering the inner surface of cavity walls with a thin stratum of material which shall subserve the purpose either of preventing recurrence of decay; or of affording support to unyielding filling material which would otherwise rest upon an unsubstantial foundation; or of preventing discoloration from filling materials liable to tarnish; or of precluding the possibility of clouding from leakage; or of strengthening frail cavity walls.

The materials used for *lining* are varnishes made from sandarac, copal, inspissated Canada balsam, mastic, etc., facing amalgam, oxy-chloride of zinc, zinc-phosphate, and gutta-percha, either in solution or as employed for filling.

Linings of varnish, facing amalgam, and the zinc plastics, particularly the zinc-chloride, I can heartily recommend, as I have used them very frequently, and for many years; but the linings of gutta-percha I can only caution against as unreliable. The solutions of gutta-percha have a worse record even than the thin linings of gutta-percha stopping, and this is needless, for a gutta-percha foundation has ever proven insecure in the extreme.

Linings of varnish are indicated in shallow cavities where only limited undercuts or retaining-holds can be obtained. They should be permitted to dry thoroughly, and will then be found preservative of tooth-bone and preventive of discoloration. R. Gum-Sandarac, grs. iij to v; alcohol, f$\mathfrak{z}$i.*

Linings of facing amalgam are indicated where contour amalgam is to be used for building portions of crowns on to remaining portions of tooth-tissue which afford but slight anchorage for the filling. In these cases no space can be given to zinc-plastics, and indeed even the thin space occupied by a coating of varnish, is better utilized by an equal thickness of filling material to which *attachment* may be made for the contouring. Facing amalgam meets this requirement.

Linings of oxy-chloride are the "regular thing." In all cases of deep undercutting; in all cases of frail, thin walls; in all cases of poor tooth-structure; in all cases of discolored

---

* Every cavity which can be kept *perfectly dry* should, on general principles, be *varnish lined*, at least.

teeth; in all cases of marked tendency to recurrence of decay, I can advise a good oxy-chloride lining.

Linings of zinc-phosphate are now becoming accepted things There are certain qualities pertaining to this material which render it more desirable, as a liner, than zinc-chloride. These are its waxy ductility in working, which permits its placing by means of burnishers instead of by pelleting; its perfect adaptation to cavity walls in such manner as to be easily and nicely brought to feather-edges at cavity margins; the reasonable strength of such edges when compared with oxychlorides; the greater celerity with which it hardens, and its valuable characteristic of non-shrinkage.*

All these give to zinc-phosphate advantages which it behooves us to utilize, and it is therefore eminently proper that experimental knowledge as to its maintenance of integrity and permanence of utility shall be obtained as promptly and as positively as is compatible with safety to such teeth as require such aid.

Mixing.— By this is meant the final uniting of two preparations, *both of which*, at the present time, *are compounds*, which mixture results in a material which is either directly used for some process connected with the filling of a cavity, or is to be utilized for this purpose by some other process, as "mixing" alloys for making another alloy.

The "mixing" of such compounds, as, by this process, make materials for capping pulps, lining cavities, or filling teeth, is always best done upon a glass-slab with a spatula. The methods of the various mixings of this kind have been given, in place, under "oxy-chloride," "oxy-sulphate," and "zinc-phosphate;" but it remains for me to say that in the making of these mixings there is the greatest possible individual difference.

From this fact it is impossible that all should attain results which shall be alike valuable or satisfactory, and it is to this, that many of the deficiencies ascribed to plastic materials are due. *Every operator* thinks that he can mix any plastic with perfect facility, whereas, the truth is that not one in an hundred can make the most advantageous "mix" with any of them. They are either mixed too thin or too thick; too slowly or too quickly; the menstruum is either overloaded or undercharged;

* See Appendix, Sec. 8.

the mix is taken for use either before the proper time, or when the setting has advanced so far as that it is interfered with, and the material rendered practically valueless in its introducing.

NOTE.—Some time since, I was requested by a manufacturer to tell him what I thought of a material which was on his office-table. I poured out upon a slab a little of the fluid and a portion of powder, and after making the mix, and noting its behavior, told him it was a good oxy-chloride. "Do you think it good?" said he. "Certainly, quite so," I replied. Upon this he showed me a letter with which the package had been returned; in it the material was stamped as perfectly worthless, and a very emphatic request made that no more of that kind should ever be sent again. After some weeks, the gentleman—who is an expert at mixing—told me that he had kept the material in his office until he had requested quite a number of visiting dentists to manipulate it, and said he, "not one made what I should call a *decent mix.*"

I could readily appreciate this, for my own office has been the scene of numerous very funny "mixes" at the hands of some very "eminent" practitioners; and it is from this want of knowledge that some of the most miserable plastics have secured "testimonials" which will in the near future be as "peculiar" as are zinc alloys, *but not nearly so satisfactory.*

Such "mix" easily; they require no knowledge; no dexterity; no promptness, neither proper patience; no manipulative ability for their using; they require nothing but "judgment."

But the "mixing" of good plastics requires everything *except judgment.* It requires thorough knowledge of all their attributes during combination; it requires, at times, exceeding dexterity; it requires, with some, a promptness which can be attained but by few, and with others a patience which is not possessed by all.

The act of using while the mix is setting requires peculiar manipulative ability, which, while, as has been said, it can be attained by a larger proportion of operators than can attain to the most excellent manipulation of gold, is, nevertheless, possessed by few, *if any,* of those whose energy has been expended in the acquirement of the ability to do elegant "gold work."

Beside the spatula *mixing* upon the glass slab, there is the "alloy mix" which is done in the "mixer" or "ager," as the instrument is variously styled, and by means of which some of the nicest working alloys are made from the several basal alloys which have been given; of these I may mention the "mix" of *contour* and *facing* as given for the making of an amalgam of remarkable characteristics for ordinary cavities in fairly good teeth; the "mix" of basal copper and basal gold front-tooth formulæ for making alloy for *front-tooth* amalgam; and the "mix" of "*coin filings*," two parts; "*submarine*," one part; for the making of a plastic-working and eminently tooth-conserving amalgam for large, shallow cavities upon buccal faces of molars of structure decidedly below medium.

There is no method, of which I am cognizant, that will so thoroughly "mix" alloys as the revolving cylinder. This should be revolved with only moderate speed, that the filings may roll together over each other, and the "mixing" should be continued for a sufficient length of time to insure completeness. This for a small lot — half a dozen ounces — requires about an hour.

Finally, we have "heat mixing," by means of which the powdered silex, feldspar and chalk are mixed with the molten wax and gutta-percha base-plate for "temporary stoppings."

Pelleting is the method by which oxy-chloride of zinc linings are best placed in apposition with the walls of cavities. It is done by first rolling several — five or six — small pellets of cotton wool, and placing the thumb-pliers and an appropriate plugging instrument in readiness for use. The lining material is then mixed in desired amount — preferably in small quantity — and is taken upon the end of the spatula and placed approximately in position; one of the cotton pellets is then taken in the thumb-pliers and with it the lining is pushed more accurately as desired; the pellet is now left in the cavity, and laying down the pliers the plugging instrument is used to compress perfectly the pellet and subjacent lining material. This pellet is now removed immediately, before the setting oxy-chloride entangles its fibres, and, *if needed*, another pellet is pressed into position. This more perfectly dries the oxy-chloride, and accurately places any portion not previously adapted. Consecutive quantities of lining are thus placed in position accurately and neatly, and in a film of such tenuity as to render its shrinkage practically nothing.

Rubbing.—This term is used with reference to mortar-work, by which is accomplished the pulverizing of ingredients, the compounding of powders, or the fusing of filings in mercury as they are rubbed together.

The "rubbing" of the *fritty* portion of the calcined mixture prior to the final compounding of the powder for oxy-chloride of zinc, and the "rubbing" of the calcined powder for oxy-sulphate of zinc, are matters upon which depend, essentially, the quality of these two materials.

If these ingredients are not *thoroughly* "rubbed," neither of the powders will permit of proper mixing; for an obdurate grit will consume so much time for its incorporation with the mass as to either deprive it of the value of a portion of material or necessitate a lengthy spatulation which injures seriously the setting qualities of the "mix."

The *rubbing* of filings into mercury, for the making of amalgam, has been specially referred to in another place. The making of amalgam is best accomplished by a combination of rubbing and palm-kneading; but of these two processes the possibility of the latter depends entirely upon the proper performance of the former. It has been taught that the filings should be incorporated with the mercury with a certain deliberation and yet with a degree of celerity; but that a fluidity of the first portion of the combined metals will alone permit of the making of a plastic mass when all the filings are in. Upon the proper plasticity of the amalgam, as the result of rubbing, depends the possibility of its subsequent *proper* kneading.

It is very easy to add a little more filings if there be too great plasticity; and it is equally easy to add a little more mercury if the mass is too hard, or even crumbly; but this is not *making* an amalgam which will "test" as would a properly made material. It will either set more quickly than it should, and will thus be less homogeneous, or it will set more slowly, and thus be less dense. Either of these conditions detracts from its edge-strength, and both influence shrinkage and bulging, and consequently crevicing. In short, a very good alloy may thus be made to make an equally ordinary amalgam.

From time to time various materials have been suggested as desirable to use in connection with the *rubbing* of the amalgam mass; some for the purpose of producing a whiter filling, or one which would better maintain its whiteness; some for the purpose of making a quicker union of the ingredients; some for the purpose of attaining greater plasticity, etc.

These additions have been such as alcohol, chloride of zinc, chloride of sodium, common chalk, spirits of ammonia, acids of varied kinds and strength, from strong sulphuric to dilute acetic, alum, borax, carbonate of soda, carbonate of ammonia, etc.

It has been directed to use these in their ordinary fluid form, or as crystals, or as dry powders, or as solutions; but the invariable final process is *the washing out of the adjunct with clear water.*

It is indisputable that some of these additions seem to produce desirable results, or rather that they produced them in connection with the amalgams which have been generally employed; but the changes in alloy formulæ, which have been given, secure results which far more than equal any obtained from such means, and the "washing" which they necessitate is, *of itself, sufficient to condemn them all.*

Setting is the word applied to the hardening of all plastic materials used for filling teeth, or in connection with the introduction of fillings, with the exception of such as contain gutta-percha; these are *said* to "*harden.*'

The *setting* of the various plastics is quite peculiar, each according to its kind; thus "submarine" amalgam *sets* with medium celerity. "Contour" amalgam *sets* with remarkable rapidity; is better *set* in ten minutes than *ordinary* amalgams are in half an hour. "Facing" amalgam is much more deliberate, and its *set* is never of that firmness and density which are so essential for edge-strength and particularly for contouring purposes.*

The *setting* of zinc-sulphate, while dependent upon its thinness or thickness of mix for promptness or deliberation, is, nevertheless, always quick after it commences. It is necessary that the material, if good, be mixed *quite thin* or *milky*, in order that it work at all desirably in capping or protecting pulps; for if anything approaching thick consistence is attempted, it will set not only before it can be used, but sometimes even before it can be well mixed.

---

* Both "copper" and "coin" amalgams require several hours, and sometimes days, before they *thoroughly set.*

The *setting* of zinc-sulphate is also peculiar in that while it sets with singular hardness, permitting a bright polish, it never loses its decidedly metallic, astringent taste, but responds markedly to a slight touch of the tongue.

It is this property which seems to give it much therapeutic value.

The *setting* of zinc-chloride is varied both by the composition and the freshness of the powder, and also by the strength and condition of the fluid. For good *setting*, which is quite deliberate for a zinc plastic, the powder should be well calcined and reasonably fresh. When, from age, the mix does not set well, it can readily be made to do so by re-calcining the powder in a crucible, or, frequently, by merely placing the unstoppered bottle of powder in an ordinarily hot oven for twenty or thirty minutes.

The *setting* of oxy-phosphate is *slow* just in proportion as the compound is poor, and is *quick* equally in the same ratio. In some conditions of fluid, syrup, or crystal, the mass retains its doughy plasticity for many minutes,—five or ten,—and is easily worked and moulded during nearly all this time. This *set* is very deceptive; for to the uninitiated it is a most pleasing and satisfactory characteristic, but with those who have witnessed the failing of fillings made from such material, the slow set is anything but desirable.

In other conditions of fluid, syrup, or crystal, the mass *sets* with instantaneous rapidity, and then, usually, crumbles into a coarse powder. This result, though proving the material EITHER BADLY MIXED or worthless, is better for the operator than is the slow set; for it compels him to mix anew, and thus definitely settle the quality of his plastic.

The *setting* of zinc-phosphate is *quick*, within reasonable bounds, just in proportion as the material is good; indeed, the *setting* of a good zinc-phosphate may be regarded as rather too quick to be considered "within reasonable bounds," for it is so apid as to preclude its *proper* working by any except dexterous manipulators.

Those who work slowly, temperamentally or habitually, can overcome or *modify* this rapidity of setting by mixing "soft;" but the *set* cement will neither be so hard nor so durable as is that which is mixed with *proper* plasticity and manipulated with *proper* dexterity.

Softening.—This term is applied equally to the preparing of filling material for filling purposes or for the rendering of filling material easy, or comparatively easy, of removal from cavities of decay.

The *softening* of gutta-percha is now done upon singly- or doubly-raised and air-protected metal plates, which method, while it does away with the inconveniences of the obsolete "water warmers," yet entails a somewhat greater caution, lest overheating and consequent spoiling of the material ensue; but the advantages of the "dry method" were so apparent as at once to command acceptance, for though there is *possibility* of this untoward result, there is no great *probability* of it.

"Low heat" gutta-percha materials may be proportionately overheated without serious detriment, but "medium heat" and "high heat" gutta-perchas should be *softened* with *much care*, as the heat which has *always* been necessary for their *softening* is more easily raised to a disintegrating degree, 240° F., which so seriously impairs the value of these excellent materials as to render them practically worthless.

The *softening* of a gutta-percha filling for the purpose of repairing, adding to, or *removal* is done by means of a heated instrument,—usually a probe, either blunt or not too fine,—with which the desired result is readily attained.

The *softening* of an amalgam filling is accomplished by drilling the greatest possible number of small drill-holes in the mass, and filling these with mercury. This is done by making VERY *soft* amalgam mass in the glass mortar, and, having taken up a portion of the make in the thumb-pliers, placing it upon the face of the filling. The mercury is then worked into the drill-holes by means of a small probe.

In a short time—thirty minutes or so—the mercury will have so united with the amalgam as to permit of its easy drilling with a *good, sharp burr-drill.* As the drilling breaks through column after column of the mercurialized amalgam, the filling will become more and more readily cut into, and will, not infrequently, break up into several pieces. In removing these *softened* fillings, it is better that the cutting in two—halving of them—should be systematically aimed at, as, this having been done, it is usually not difficult to gently force the separated halves from the sides of the cavity, and thus take them away.

Shrinkage.— By common consent and habitual usage, the contraction of "plastics" is called *shrinking;* the relative amount of contraction, their "degree of shrinkage;" the instruments for determining this, "shrinkers;" and the other methods for attaining the same results, "shrinkage tests."

For ascertaining the *shrinkage* of amalgams, the three methods which have been adopted as reliable are, the "index micrometer," the "tube test," and micrometric observation, in regard to each of which details have been given.

By these means great advance has been made in diminishing the degree of shrinkage in all the approved amalgams, while, at the same time, they have been steadily gaining in the varied attributes of strength of edge, rapidity of setting, maintenance of color, plasticity, and compatibility with tooth-tissue.

By micrometric observation it is shown that *shrinkage* of good amalgam has been reduced to less than 1–1000th of an inch in an ingot of two micrometric inches. This would give as the *shrinkage* of a quarter of an inch filling the very small degree of the $\frac{1}{16000}$ of an inch.

For proving the shrinkage of gutta-percha, it is usual to employ the "tube test," in which an uncovered tube — small vial, made of extra thick glass tubing — is carefully packed, under visual scrutiny, with the preparation to be sampled, after which aniline ink is poured into the vial. In a short time the shrinkage of the gutta-percha is proven by the permeation of the purple ink between the filling and the glass.

Ivory tubes and cups have been used for this purpose, as imitating closely the actual tooth; in these the shrinkage of gutta-percha — both red and white — is demonstrated to be something quite notable, as leakage takes place very freely.

The *shrinkage* of oxy-chloride of zinc is well demonstrated by means of a small ring of glass tubing — say one-half or three-quarters of an inch in diameter. This being packed as for filling, and allowed to stand for a few days — three to five — will show *shrinkage* in quantity.

The *shrinkage* of zinc-phosphate is tested in the same manner as is that of zinc-chloride, and is shown to be merely nominal.

Tapping.— This word, used in dental therapeutics with reference to entering pulp cavities other than through cavities of decay, is employed in plastic dentistry with reference to the packing of amalgam.

For many years it has been directed to pack amalgam by rubbing it *against* the walls of cavities, and *upon* such portion of amalgam as is already introduced. It is most conclusively demonstrated, by tube packing, that this is a very unsatisfactory and incomplete method of inserting amalgam. It is therefore directed that it be done by "tapping" the first pieces into apposition with the cavity walls, and that the consecutively introduced pieces be made to unite, homogeneously, with those previously introduced, by "tapping."

Tapping consists in delivering light blows, from the appropriate filling instruments, upon the amalgam *after it has been crushed* into approximate position and apposition. This "tapping" is *not to be done* with mallets, either hand, automatic, or electric, as a different kind of blow from any so given is far preferable. The "tap" from the filling instrument — the same used for crushing — is a mingled push and blow, which is soon acquired, and is as promptly recognized as very efficient in producing admirable results.

Testing.— This word is used in reference to an immense line of work which has been required by the exigencies of development.

It is such work as has been but little demanded in connection with the use of gold, for the "testing" of this material is confined to its manipulation during insertion, while the "tests" which have been applied to it meet with no concurrence from a plastic-filler as to their value. The *amount* of any material packed in a cavity — decided by its weight — is held by the plastic-filler as of little, *if any*, import as regards the tooth-saving value of the filling.

The *solidity* of any filling, *even of gold*, is regarded by the plastic-filler as *no criterion* of the real, tooth-saving worth of the work.

Even the *apposition* of the filling material to the walls of the cavity is viewed by the plastic-filler from his peculiar standpoint; for he notes the constant failure of the gold fillings with their elegant microscopic adaptation, and the wonderful preservation afforded by gutta-percha with its known deficiency in that respect; thus it is that the circumscribed line of work in the "test" packing of gold, surrounded as it is with the high-sounding epithets of "artistic," "elaborate," "ideal," etc., is viewed by the plastic-worker with mingled wonder and amusement, and is thought to have but little "weight," less "solidity," and no appreciable "apposition" to the great question of tooth-salvation.

The "test work" of plastics is the natural result of a multiplicity of materials and of the desired modifications in connection with numerous compounds. Not alone have the attributes of silver, tin, gold, copper, zinc, mercury, gutta-percha, oxide of zinc, sulphate of zinc, chloride of zinc, nitrate of zinc, borate of soda, the acids of phosphorus, alcohol, the gums of mastic, sandarac, copal, caoutchouc, and acacia, oxide of tin, wax, sulphite of lime, feldspar, and silex to be considered, but the endless variety of composition in which these ingredients may be advantageously employed; and when to these are added the "questionables," of doubtful value, but which have claims that entitle them to attention, such as antimony, bismuth, cadmium, palladium, alumina, gypsum, chalk, alum, kaolin, and quick-lime, is it strange that the word "testing" has a meaning for the worker in plastics beside which the repetitions that have formed the "discussions" of dental societies for the past twenty years seem like time-wasting trifling?

On the one hand, we have the rehash of methods and manipulations which have resulted in making it questionable whether the title of "permanent denture" is most appropriately applied to natural or artificial teeth! while on the other, we have that work which by "*testing*" has resulted in materials, methods, and manipulations which fairly bid defiance to the ravages of caries, and make of the frailest and softest teeth *wonderfully* "*permanent dentures.*"

In a paper which was read before the American Dental Association, Niagara Falls, August, 1878,—*not published in the Transactions*,—reference was made to the statement, from one of those to whom students of dentistry look for instruction, that plastic work was "*guess work.*" To this I replied that "so far from being '*guess work,*' we have a range of 'tests' which tell us very well what we may expect of any material under given conditions. These are such as strength test, edge test, setting test, shrinkage test, expansion test, color test, heat tests (wet and dry), leakage test, frotting test (for probable wear), acid test, alkali test, conduction test (electrical and thermal), and finally the oral test, which decides the compatibility of materials with tooth-bone and their behavior *in the oral fluids* and *under oral influences.*

"*By means of these tests we are enabled to make a choice of material to meet the varied indications that constantly present in practice, which, to our apprehension, approaches to* SOMETHING LIKE SCIENCE.

"It is by these means that we frequently combine two, three, four, or more different materials in the filling of one cavity, each of which best subserves its purpose in its appropriate position, and insures an operation which for comfort, beauty, and permanency can *in no other way* be equalled."

All this is so entirely at variance with that stereotyped practice which it is *the duty* of the publishing committee to place, *year after year,* before the profession, that it seems "reasonable and consistent" it should be deemed *imperative* to *suppress it.*

This alone would doubtless have been sufficient to decide the action of any dental publication committee, but when to it was added the assertion that Plastic Dentistry "is *even now* a very well worked-up *specialty* of our profession; one that is *so little known as scarcely to be mentioned* except with contempt or disapprobation from the *lecture stand,* AND YET *one which can now take any denture so forlorn as to be* HOPELESSLY *abandoned by the* BEST GOLD OPERATOR *in the world* and make of it a COMFORTABLE, SATISFACTORY, and BEAUTIFUL SUCCESS," then *suppression* of such a damaging assertion, particularly if it was in the least degree probable that it could be sustained, became a *duty* which, in its magnitude, was simply overwhelming.

What an extraordinary thing is *the sense of duty.**

I have felt it *my duty* to place these statements before my professional brethren *because* they have all been made as the

---

* Refer to letter of "Publication Committee" of American Dental Association, p. 61.

results of long and thorough "*testing;*" because I am conversant with the continued, persistent work which has been required to do this; because I know of the earnest interest with which the work has been done; because I know of the thousands of patients who have *realized* the benefits of the work; because I know of what I speak — that it is capable of demonstration that it is *true;* because I feel that my profession *needs* it; because I feel that suffering humanity *needs* it; because I feel that the cause of *progress* demands it.

Thus it is that in *my* "decision" I *am* "actuated by personal prejudice." I *do consider* the statements of "sufficient value" to warrant their presentation; I *am* "influenced by" most decided "opposition to the doctrines advanced" in the usual dental contributions; and I also am "governed solely by *a sense of duty.*"

Trimming refers to the cutting away or removing of surplus filling material. In plastic filling, it is almost universally the case that a superabundance of material is introduced. This is due to the fact that no more time, if as much, is required to more than fill any cavity than would be needed for the accurate filling of the same. It is also found to be advantageous that an ample amount of filling be given for final shaping. It is less difficult to remove surplus than it is to add on. The materials are comparatively inexpensive, and thus "waste" does not involve much "loss." The removal of surplus does not necessitate inflictive or disagreeable instrumentation, such as hand-filing, chiselling, stoning, etc., as is the case with gold.

In *trimming* amalgam fillings, all grades of tool-edge are not only admissible but desirable, from the sharp edge of the "trimmer" to the absolutely round edge of the curved burnisher. This latter is particularly useful and non-inflictive for trimming the fillings and making smooth adaptation of filling with cavity edge below the gum, especially on buccal faces; while the thin-edged, paper-like separators are admirably calculated for trimming between teeth.

For *trimming* fillings of gutta-percha, I use exclusively heated instruments, several of which being heated at once upon the "tool-heater," it is easy to adapt varied shapes to varied requirements.

For *trimming* zinc-phosphate fillings, sharp-cutting instruments are required, such as knife-edged trimmers, burrs, and files. These should be kept perfectly dry or thoroughly wet, as by these means the filings or cuttings are either dry powder, which can be blown away, or make, by free dilution, a milky fluid which does not seriously clog either burr or file.

Trunnioning.—It is by this operation that amalgam buffers of all sizes are retained firmly in position, when they do double duty in that they prevent wear from attrition upon gutta-percha guards, and disintegration, by fluids of the mouth, of oxy-chloride linings. Suppose a cavity in a molar or bicuspid, mesial or distal, reaching up to or under the gum, nearly into the pulp, extending partially over the articulating face of the tooth, and having very frail, thin walls both buccally and lingually. This condition is *frequently* found in bicuspids of soft structure which have been filled *repeatedly* with gold. In such the cervical edge is first guarded by a layer of gutta-percha, which is neatly extended far enough into the cavity to prevent pulp irritation from conduction after filling, and from zinc-chloride irritation during lining. The cavity is now lined buccally and lingually with oxy-chloride of zinc, or zinc-phosphate, which is allowed to set thoroughly, covering it temporarily with "temporary stopping" for an hour or two, or a day or so if more convenient.

When the zinc lining is perfectly set, an indentation or pit is drilled into each side of it, buccal and lingual, as large as is consistent with strength; a contour amalgam filling is now made, which by entering the drill-pits becomes held by *trunnions*, and thus solidly maintains its position.

Wafering.—Upon the intelligent utilizing of this process depends largely the comfort, satisfaction, and success of amalgam work. It consists of making *wafers* of small portions of amalgam, and using them for the purpose of increasing the consistence of amalgam which has been introduced *soft* for some specific object, as prevention of pulp irritation, prevention of too severe impinging upon exquisitely sensitive or heavily decalcified dentine, or, to facilitate its insertion into some exceedingly inaccessible location; *or* to increase normal consistence of already introduced amalgam for the purpose of hastening setting, and thus permitting prompt continuance of work, as in building; *or* for the hardening of the face of a filling, that the work of finishing may be facilitated, while at the same time an increase of edge-strength and a whiter face may be insured.

The *wafer* is made by taking a small portion of the amalgam and folding it in a piece of "chamois-skin;" the chamois is now twisted so as to secure the amalgam and prevent such escape of mercury from squeezing, as may permit of its again being taken up from contiguity by the wafer.

The squeezing of the amalgam should be done with large, strong pliers — as illustrated — made expressly for this purpose; these are about seven inches in length, with flat, round-edged jaws — so as not to cut the chamois-skin — and are properly adapted to *thorough* squeezing with the expenditure of comparatively little force.

This is an important consideration, for, with ordinary sized pliers, either the squeezing would be insufficient, or the force required would be inflictive to the hand. The difference in the physical characteristics of *wafers* made by using the ordinary pliers and those made by the *appropriate* pliers can only be realized by comparison of the two results, and the success of *wafering* is widely different when done by *thoroughly-squeezed* instead of *half-squeezed* "wafers."

Washing.— I have merely introduced this term that I may once more enter a protest against the process to which it refers. Every experiment points to the inutility of "washing" amalgam; every experiment indicates that it is worse than useless; that it is detrimental. It is conceded that fillings made with submarine amalgam would, theoretically, be better made if done dry; while the fact that good results follow *wet* work is only urged in the humane effort to accomplish respectable tooth-saving without the necessity for dire infliction. Therefore, I should condemn any process, in connection with the making of amalgam, which entailed, as a part of its performance, the *washing* of the material.*

Weighing.— In plastic filling this word refers only to the manner of determining the relative quantities of mercury and filings required for the making of any definite amalgam; or for the modifying of the known *best* make of any given amalgam that exigencies may be accurately met.

---

* Except for "copper amalgam" made from dry precipitated copper. See p. 104g.

I think it sufficient to prove the importance which I attach to the *weighing* of the ingredients for making amalgam, when I state that, with the daily experience which I have had for nearly thirty-five years, *I never make amalgam without weighing the proportions.*

Recognizing completely the *necessity* for accurate compounding in the obtaining of definite results, I *know* that *I* cannot *even closely approximate* the desired amounts of mercury and filings, and I further know that *if I could* it would not be sufficiently accurate for my purpose.

Upon very many occasions, I have requested, from those who claimed ability to "judge" with "reasonable accuracy" in this apportionment, demonstrations of their skill, and I *think* the sequence has been, almost universally, a determination upon the part of the experimenters to *weigh* their proportions in future.

The fact is, *no one* can secure anything more than a very rude approach to the required amounts of mercury and filings for the making of amalgam, if this is done by the "judgment" method; whereas, by the *weighing* method, one can soon become so expert as to prognose the consistence or plasticity which will pertain to nearly every "make" of amalgam mass *before it is rubbed.*

This is what is required; for thus it is that the proper plasticity for meeting special indications is given any appropriate selection of amalgam; thus, a submarine mass, which is to be used in an accessible cavity upon the buccal face of a lower first molar, would naturally be made of firm, workable consistence; while the same alloy would, as appropriately, be made into an amalgam of *soft plasticity*, if the cavity to be filled were exceedingly inaccessible, and was possessed of frail walls, spottings of excessively sensitive dentine, or other like complications.

By *weighing*, all these important considerations are promptly, neatly, and accurately met; and I think that years of practice will only serve to demonstrate to others that which has been demonstrated to me, viz., *by no other method than* WEIGHING, *can satisfactory making of amalgam be accomplished.*

I will repeat, that the *weighing* is not to be done with weights, but by weight; and I would say, that although the making of amalgam from reliable alloys is *always* done with an *approximately* horizontal scale beam, yet, to the expert, the slight deviations from this are the means by which he uniformly accomplishes any desired modification.

Whitening.— This term is applied alike to teeth and to fillings. In plastic dentistry, the "*bleaching*" of teeth is ignored as a detrimental and non-compensating process, and consequently the word has become obsolete.

In the article on oxy-chloride of zinc, its use as a *whitener* of teeth has been explained, and reference has been made to the permanence which attaches to this restoration of color.*

In the article on "insertion of amalgam," the peculiar method of *white finishing* has been described; but as this was then only one point among many, and as considerations of *beauty* have obtained notable prominence in "plastic-work" ever since reaching conclusions upon the *practical* problem of *tooth-salvation*, I regard it as proper to direct attention again to the finishing of amalgam fillings.

After the required ten or fifteen minutes of setting, an amalgam filling may receive, if needed, its accurate contouring by trimming, and should then have its second finishing from the piece of pine stick as described in the article referred to.

This finishing results in a white face, which, though yet of comparatively coarse grain, is nevertheless much finer in appearance than was the grain given by the first smoothing. Again, ten minutes or so should be given for further setting, when the filling may be finally finished. The final finish is given by the soft pine stick, but it is sometimes found necessary to add to this a little *finely levigated* pumice. This is made by putting in a basin part filled with water a small quantity of such pulverized pumice as is usually sold. It is then thoroughly stirred up, and allowed to settle perfectly. The clear water, with the floating impurities, is then poured off. The basin is again partially filled with water, and the pumice again thoroughly stirred up. This is then allowed to settle for a *few seconds* — four or five — when the *milky fluid* is poured off, and the fine pumice contained in it is permitted to settle until the

---

* The *whitest* lining for restoration of color is made by mixing *clear oxide of zinc* with *either* of the fluids of oxy-chloride or zinc-phosphate, as indicated.

water is again perfectly clear. The clear water is poured off, and the remaining fine sediment is dried for use.

When the final finish is given, it will be noted that the most decided *whitening* is accomplished by smoothing the face of the filling in a *downward* or *upward* direction,—as it is in an upper or lower tooth,—and not across the filling disto-mesially.

This method of finishing leaves a striated surface, which, though microscopic in the fineness of its markings, yet presents a succession of faces upon which the light impinges directly, and by which a reflected *whiteness* is given to the filling.

CONCLUSION TO FIRST EDITION.

WITH this I am permitted to see the end of a work which was commenced a quarter of a century ago.* Its possibilities were then discussed, its probabilities foretold, but the view of its realization—then, to a few, an ideal of the far-off future—has been granted to me alone.

Although it is with grateful satisfaction that I now contemplate the gradual development of that which has been the medium for so much alleviation to suffering, so much bestowal of long-enduring comfort, and such incontestably beneficial results, it is yet with mingling of regret and pleasure that I cast it forth to do battle in the struggle for professional place and precedence.

I shall regret to see it frowned upon; I shall regret to see it misrepresented, and yet I can but know that it must meet with the common reception of those numerous predecessors which have waged war against "accepted doctrines;" but it is to me a pleasure indeed that I am, even yet, afforded ample opportunity for the practical demonstration of the truths of its teachings, and that strength is yet given me to raise my voice in its defence.

I ask for it a serious, thoughtful consideration, and sincerely hope that much good may come of it.

* 1855.

APPENDIX.

Section 1. Refer to pp. 65, 176.—A valuable suggestion has been made by Dr. J. F. Siddall, which consists of "Heat Ageing" freshly-cut alloys. This is done by applying a moderate degree of heat to the filings for the space of three to five minutes. Experiments have proved that the best method for doing this is that of placing the filings in any convenient receptacle, and subjecting them to the heat of boiling water. By this device one is enabled to use freshly-cut filings with much satisfaction; but I have found that they will yet gradually improve by "time ageing," even after having been subjected to this process. An excessive degree of heat will render fresh filings of "excellent" quality almost completely devoid of "setting" power, and it is therefore important that the security of the water-bath should be utilized in prevention of this possible result.

Section 2. Refer to page 67.—A very desirable form of mercury holder has been invented by Dr. J. H. Kidder, and has been thought well worthy of illustration (see opposite page 90), as it has proved markedly economical.

By a gentle turning of the upper section of the holder, a quantity of mercury exactly proportioned to any given requirement is forced through a minute orifice into the inverted cone-shaped receptacle, and from this is emptied into the scale-plate.

This ensures the use of only an equally just amount of filings for each "make" of amalgam, and the saving which is thus accomplished is really a matter of much moment. I have found the *simplest* form of this instrument the *best*, and that which has the orifice at the *bottom of the cone* much preferable to that which has the orifice upon one side.

Section 3. Refer to page 90.—In writing of "ground glass" mortars, I have reason to fear that the usual idea which is attached to the term "ground glass" has overshadowed the statement, "but if the glazed surface is *delicately* taken off the inside of the glass mortar, it will prove much superior to porcelain." I therefore desire to impress the fact that mortars should not be "ground," in the ordinary acceptation of that term as applied to glass, but that only the smooth and polished glaze should be "delicately" removed from the mortar by means of a solution of hydro-fluoric acid, having the upper half of the inside of the mortar protected by glass-etchers' black wax.

The pestle should be left polished, as, while making the amalgam equally well, adhesive particles are more readily removed.

SECTION 4. Refer to pp. 67, 88.—Although alloy metals, in notable quantity, are taken from amalgam when surplus mercury is removed by "finger squeezing," it has been positively determined that no change occurs in the relative proportions of metals composing alloys when the surplus mercury is expressed through "chamois skin."

Following the work of several others, I tested the mercury thus expressed from *one hundred* makes of amalgam, finding only "a trace" of other metal. This experiment was made three times with like results.

SECTION 5. Refer to page 167.—While it is productive of a better zinc-phosphate cement if the mass is taken from the spatula and gently kneaded, it is nevertheless advisable that it be occasionally used when only spatula mixed. This is especially the case when "facing" is to be done and when "linings" are to be made. In "facing," the mass should be well made and placed in position from the spatula,—No. 12 of the set of instruments, page 93,—and then worked into place and pressed into contour by instruments 6, 7, 8, oiled. In "lining," cements made from oxide of zinc and the fluid of zinc-phosphate, or a mixture of oxide of zinc and nitrate of zinc with the zinc-phosphate fluid, will be found to work very nicely, producing good shades, retaining their plasticity desirably, and hardening sufficiently for strength and durability with reasonable celerity.

SECTION 6. Refer to page 127.—One of the most remarkable contributions to operative dentistry is that of Dr. C. H. Land in his variously and most ingeniously devised combinations of porcelain with zinc-plastics and amalgam; and no other incidental to this work seems to promise greater range of utility than that of fire-gilding appropriate portions of fillings and *crowns*. By this means, supplemented with the amalgamating of the gold face, a possible union of fillings to amalgam "linings," and of crowns to amalgam root fillings, is attainable by "cold soldering," which is marvellous in its strength, and which bids fair to revolutionize all preconceived ideas in connection with crown work.

SECTION 7. Refer to page 166.—Yet another use for zinc-phosphate is found in connection with the unique operation of "jacketing" those deformities known as "peg teeth."

Probably no other peculiarity gives such abnormality of expression to an otherwise perfect denture as does the presence of one or more of these pointed teeth.

Until now the only remedies for this condition were extraction and replacement by transplantation of a normally shaped tooth; or by the insertion of an artificial tooth, either by plate or by filling attachment to adjoining teeth; or by pulp devitalization, crown excision and utilization of the almost always equally abnormal root for the support of a porcelain crown.

Now, by one of the devices referred to in Section 6, a platinum "jacket" is fitted to the deformed tooth, a selected porcelain facing is fused to the "jacket," and the whole is secured to the "peg" by means of zinc-phosphate.

Subservience to this requirement is another evidence of the singular utility of this remarkable plastic in response to special demand.

With its combined characteristics of adhesiveness, maintenance of form integrity, moderate continuance of plasticity, sufficient promptness in setting, sufficient strength, freedom from discoloration and comparative insolubility when reasonably protected from the fluids of the mouth, it ranks as the best of the very few cements which are at all useful for the obtaining of this very important and very beautiful result.

SECTION 8. Refer to page 191.—Linings of both zinc-chloride and zinc-phosphate are made, when indicated, of *creamy softness*, and, being followed by *immediate* introduction of filling, they are not only made exceedingly thin but also give additional guarantee for retention of fillings.

CONCLUSION TO THIRD EDITION.

TEN years have passed since it was permitted that I should publish the results of a life-work devoted to Plastic Dentistry.

In the nature of things this work was largely done from an *empirical* standpoint, and therefore from under a professional clouding.

The course of events gave to dentistry the Palmer theory of the "compatibility" or "incompatibility" of filling material with toothbone, and thus afforded a *scientific basis* upon which to found a practice which had already been proven eminently satisfactory.

This union of Theory and Practice gave birth to an organization called "The New Departure Corps," composed of Prof. Henry Morton and Prof. M. B. Snyder, scientists; Messrs. Jacob B. Eckfeldt and Patterson Du Bois, assayers of the Philadelphia Mint, metallurgists; and Drs. S. B. Palmer, Henry S. Chase and J. Foster Flagg, dentists.

The work of this organization, in turn, gave to our profession a "creed," of which, as opposed to the prevailing views, the *Dental Cosmos* said, editorially: "Both theory and practice are so essentially, so radically, so diametrically in opposition to the theory and practice which have so generally obtained, that it is quite permissible to designate them by way of distinction as the 'accepted' creed and the 'new departure' creed."

Upon this "New Departure Creed" rests the whole teaching— theoretic, practical, experimental and manipulative—of this work upon "Plastics and Plastic Filling."

Recognizing that the influence which has thus been exerted upon the practice of dentistry is indisputable, and that the old-time methods and means have been graphically and truthfully spoken of as "broken!" it seems that I could not better conclude this effort than by republishing the creed as given in 1877, and supplementing it with an appeal for its careful scrutiny and studious consideration.

THE NEW DEPARTURE CREED.

I. In proportion as teeth *need* saving, gold is the *worst* material to use.

II. Neither "contouring filling" nor "separating teeth" has much to do with the arrest of decay.

III. Failure in operations is mainly due to incompatibility of filling material with toothbone.

IV. A tooth that can be so treated as to be satisfactorily filled with *anything* is worth filling.

V. Skillful and scrupulous dentists fill with tin covered with gold, *thereby* preventing decay, pulpitis, death of the pulp, and abscess, and *thereby saving the tooth.**

VI. A filling may be the *best known* for the tooth and yet *leak badly*.

VII. Gutta-percha *properly used* is the *most permanent* filling material we possess.

VIII. A *poor* gutta-percha filling *in its proper place* is better than a *good* gold one.

IX. Amalgam *per se* is an *excellent* filling material.

X. The use of "plastic" filling materials tends to lower that dentistry which has for its standard of excellence "ability to make gold fillings," but very much extends the sphere of usefulness of that dentistry which has for its standard of excellence "ability to save teeth."

Frequent erroneous quotation, persistent misrepresentation, and innocent and ignorant misinterpretation of the *spirit* of this creed have failed in preventing its constant presentation, its unwavering advocacy and its persistent practical demonstration, and to these must be ascribed its gradually growing acceptance and the inevitable marked change in the work of operative dentistry.

My many years of practice, my long and carefully tabulated series of observations upon dental work, and my close connection with both gold and plastic efforts for the saving of teeth, have led me to the conviction that *the high style, scientific use of plastics is the culmination of dental effort*, and that while the use of *gold* for the filling of easy accessible places in good, strong teeth stamps it as the *king of all filling materials*, because its results, where it is indicated, are *everything* that could be desired, it nevertheless remains true that every *real necessity* in the whole range of dental trouble is *best* met with plastics, and

---

* This article was introduced in the creed because it was *textual authority*, at that time, that "unskillful and unscrupulous dentists fill with tin covered with gold, *thereby* causing galvanic action, pulpitis, death of the pulp, abscess, and *loss of the tooth*."

CONCLUSION. 213

that this is shown in every case in just proportion with the *knowledge* which the operator possesses in regard to "plastic work."

For this reason it seems to me that this work is fast taking not only a prominent position in dental science, but that it is rapidly gaining for itself a *position of first importance*.

To-day the educated plastic filler can do, *in large proportion*, everything that the gold filler can do; and to-day the gold filler can do only a very small proportion of all that the plastic filler can do.

The broad work of our profession goes on: in mechanics, in ceramics, in surgery, in pathology, in therapeutics, the strides are such as to require, *absolutely*, a superhuman effort to keep abreast of all! And yet with all this, I feel to urge that *the* work of the near future, *the* work for those in the early prime of practice, and *especially for those now about entering upon practice*, is most largely found in the study and development of "Plastics."

It must be known and recognized that but few indeed of this later generation had other than a most superficial knowledge of this subject.

It must be recognized that those who have most loudly decried both methods and materials pertaining to *Plastic Dentistry* know practically *nothing* of either; that their efforts at utilizing them, from the selection of the different materials to their preparation and introduction, were amusing and amazing in the extreme; that the reports of Society meetings, with the remarks of the "eminent" (as they are called), show palpably an *eminent* ignorance which it is difficult to imagine possible; and that it is only those who, forming the anomalous "rear-guard" of the past and "advance-guard" of the future, have deemed it worth their while to inform themselves, in somewise creditably, upon this subject.

It should further be known, that even of those most proficient in this regard among the active workers, few have any more than what might justly be called superficial education as relates to plastic materials and plastic work.

Few are able to discriminate between *very good* and *very poor* materials, until they have *bought* and *tried them* in the very expensive way of testing them *in the mouth*, doing hundreds of dollars' worth of work with materials which by the initiated would be at once discarded as worthless, and doing this by methods which would as readily be known as futile.

Few know of the components, to say nothing of their proportions, of which are made the plastic materials they daily use; fewer yet know anything of the various methods of manufacture, and thus are easily deceived by every foolish claim of every unprincipled advertiser.

Few know how to use good plastics to good advantage, and very, very few, as yet, know how to use them in the best known way. This is demonstrated at every Society meeting where this work is tried. Scores are willing *and anxious* to show their skill with foil, but few are they who *dare* to show their skill with plastics?

And is this because there is no *skill* required? Is it because there is no *knowledge* to be demonstrated? Let each one ask himself if he believes that it is so.

And yet the last resort in every case before extraction is plastics.

Why is this? Is it that by some mysterious power, no matter how used, no matter against what odds, no matter how grave the emergency, no matter how hopeless the effort, that *plastics* are yet *worth* trying?

After everything else had failed, would even the best and grandest of all the manipulators *try gold?*

Then why is it that, after the *best* manipulators have failed with gold, the ordinary manipulators *try plastics?*

The answer to this is given in few words, but those few words are worthy of *most serious, earnest thought. It is because, in many instances, the latter have succeeded where the former have failed.*

Had the latter failed time after time, and more, had they failed *universally*, is it reasonable to suppose that such work would steadily *and increasingly* go on?

And, on the other hand, if the efforts with plastics in such cases were *ever* successful, would it not be a subject for deep thought and thorough investigation?

How much more, then, should careful attention be given to the fact that *such* efforts succeed in *many* cases.

And finally, if such efforts succeed in many cases, what study is more imperative for coming men than that of thorough preparation for the *proper* use of those methods and materials which, imperfectly used, are capable of doing so much good?

Like every other study, that of "Plastics" opens wider with every forward step. Its breadth and depth are no more realized in dentistry than are the wonders of mineralogy by those who crack the stone ballast for our railways!

It is because of this belief, and because of the possibilities I recognize as the result of all these years of labor, experiment and observation, that I now thus present this subject; that I urge the fact that "Plastics," as I have said, *are rapidly gaining for themselves a position of first importance;* that upon knowledge of this subject will largely depend the capability of the dentists of the next generation—*the generation now upon us*—to cope with conditions they will have to care for; that upon knowledge of this subject will largely depend the position which each one will hold among his professional brethren and in the community in which he lives, *and that upon knowledge of this subject will largely depend the* ABUNDANCE *of relief from suffering and maintenance of health and comfort which dentists will be enabled to bestow upon their fellow-men.*

www.ingramcontent.com/pod-product-compliance
Lightning Source LLC
Chambersburg PA
CBHW032149230426
43672CB00011B/2498